GOD AND
CONTEMPORARY
SCIENCE

To George Shields,
with gratitude for your
fine hospitality, in recognition of a wealth of
common interests, + in
hopes of future collabora-
tion...

Philip Clayton
April 10, 1999

EDINBURGH STUDIES IN CONSTRUCTIVE THEOLOGY

GOD AND CONTEMPORARY SCIENCE

———∿∿∿∿∿⋒⊙⋒∿∿∿∿∿———

PHILIP D. CLAYTON

WM. B. EERDMANS PUBLISHING COMPANY
GRAND RAPIDS, MICHIGAN

© Philip D. Clayton, 1997

Edinburgh University Press
22 George Square, Edinburgh

Typeset in Bembo
by Pioneer Associates Ltd, Perthshire
Printed and bound in Great Britain by
Cromwell Press, Trowbridge, Wilts

ISBN 0-8028-4460-X

This edition published in the United States of America
through special arrangement with Edinburgh University Press
by Wm. B. Eerdmans Publishing Co.
255 Jefferson Ave. S.E., Grand Rapids, Michigan 49503

CONTENTS

—·ᴧᴧᴧᴧᴧᴧ@ᴧᴧᴧᴧ·—

CONTENTS

PREFACE

It has been my goal to write a book in theology that would be helpful both to readers who identify with Christian belief and to those who do not hold this belief. It is thus a book about Christian theology not written for insiders alone. Those with an interest in the subject matter are welcome on every page – and I hope they will be my most astute critics of every argument.

Consequently, I do not ask for a crucifixion of the intellect at the door; philosophers are not required to shut down their finely tuned sense for good and bad arguments in order to walk this terrain. I do presuppose some interest in the beliefs that set Christians apart from, say, Buddhists; but appropriate interest must be presupposed in *every* area of academics and science. I have also talked without embarrassment of Christian doctrines, scriptures, beliefs, practices, most of which cannot be given a foundation through any universal philosophical arguments. Recognising the particularity of Christian beliefs is not the same as erecting a barrier to exclude those beliefs from scrutiny. Doctrines are not 'dogmas' in the sense of intolerant assertions aimed to condemn those who believe otherwise; each one is meant to be an inducement that encourages reflection rather than bans it. Likewise, the word 'doctrine' does not mean 'that which is above all criticism'; 'dogmas' do not refer to implausible beliefs imposed by the authority of tradition upon incredulous readers; and 'religious faith' does not just mean 'that which is opposed to human reason'.

Instead, I shall use 'doctrine' to mean formulations of beliefs that seem to be implications (or outworkings) of the fundamental belief that Jesus was involved in some special way in God's action and self-revelation in human history. Christian doctrines are the theorems – even *hypotheses* – that aim to express what basic or 'minimal' Christian belief might mean and entail. If at any point in what follows I have thought badly about the 'logic' of Christian doctrine, I wish to be corrected and will retract the appropriate claims. For such thinking is a public matter. It may well be that non-Christian readers will pick up inconsistencies more quickly and astutely than Christian readers will – those who move comfortably, sometimes too comfortably, within the orbit of these beliefs and terms. One does not have to accept the starting point, the revelatory significance of that person Jesus and his actions, to criticise what I claim follows from it. But one *does* have to grant me this starting point at least *ex hypothesi* – for without this set of assumptions Christian theology would be merely the philosophy of religion done too narrowly. Of course, I *hope* that the systematic thinking through of my starting point and its implications will tend to make this perspective more plausible for contemporary readers, whether they begin as atheists, as agnostics, as Christians or as participants in other religious traditions.

And yet . . . Christian theology is also done in the service of the church – in divinity school classrooms, in seminary and Bible school classes, for those in the pulpit and those in the pews, theology for those who hold Christian beliefs and engage in Christian practice. When Rabbi Jonathan Slater speaks in my classes, he presents Jewish theology to (mostly) non-Jews with humour and intelligence, and yet he studies Jewish thought in the first place in order to minister to Jews. I trust the intellectual credibility of the following pages will not be compromised by the fact that they are written by a Christian thinker with the goal of helping Christians (and other theists) to think systematically about a burning problem for faith: the problem of God's activity in the world in an age dominated – even for believers! – by scientific assumptions.

I have said less than might have been said about another perplexing question for Christians: the person and redemptive role of Jesus Christ. Here the challenge is not science so much as the plurality of the world religious traditions, with their competing claims about what salvation is and how it is achieved or pursued. I have also devoted too few pages to the nature of the church and the Christian eschatological hope. Fortunately, theology is a cooperative enterprise; other books in the present series pay detailed attention to topics I can only touch in passing. No inferences should be drawn about the importance of other topics in

Christian theology from the fact that I was not able to cover them fully here.

Much of the discussion involves a three-way interaction between theology, science and philosophy. The nature of the present series has not allowed me to delve as deeply as I would have wished into the scientific data and its competing interpretations. It should be clear to the reader that to carry out the task in full adequacy would require detailed discussions of (seemingly) purely scientific matters. To be specific, if there were space enough and time, a theology of the God/world relation as I envision it would have required a slow, careful working through of physical cosmology – say, on the scale of Willem Drees's *Beyond the Big Bang* – and of the various stages by which the biological and psychological realms emerged. Regrettably, that task could not be completed here. Happily, however, I can refer the reader to a book in which this task *is* carried out, a book referred to often as I wrote these pages: Arthur Peacocke's *Theology for a Scientific Age*, especially Part I.[1] I have presupposed Peacocke's excellent summary of the data in what follows and have not tried to duplicate it. Astute readers will note that the present book, like Peacocke's, defends a panentheistic doctrine of God. Because of the extent of our agreement, the two works can probably be read as complementary, the one more technical on the science side, the other more technical on the philosophical/conceptual side.

The Christian scriptures, to which I defer here, use the male pronoun in referring to God, although theologically, of course, God is neither male nor female. I have done the same here, and with the same intention. Soon, perhaps, an adequate solution to the pronominal problem will be found. To counteract in a small way the prejudicial effects of the use of the traditional pronoun, I have used the feminine pronoun for all generic references in this book. Unless otherwise indicated, translations of foreign-language texts are mine and parenthetical page references refer to the work cited in the previous endnote.

It has been my pleasure to work with five sharp research assistants during the writing of this book. Jane Beal, Summer Jackson and Dina Laumann provided excellent research support during the writing process. Tom Beasley deserves thanks for his careful work in preparing the index. Kerry Kovarik's own thinking grew in pace with the book, until at the end he became as much discussion partner as research assistant. I am grateful to the California State University system, and to my colleagues at Sonoma State, for a one-semester sabbatical leave during which the final drafts were completed, and also to Tricia Lewis, who typed large portions of the manuscript with unflagging accuracy. At Edinburgh University Press, Jane Feore as Acquisitions Editor and Peter Williams as copy-editor provided professional guidance and corrected a multitude of sins.

As usual with any long-term writing project, I owe an immeasurable amount to my colleagues and discussion partners around the world. To many who remain unnamed I have nonetheless paid the highest compliment: I have been convinced by their arguments, as they will discover in this pages. Four special debts beg, however, to be acknowledged. Wolfhart Pannenberg read and commented on the project, both before and during the writing phase, with his usual sharpness and critical acumen. As a theologian I have learned more from Pannenberg's critical guidance than from any other, and it is a pleasure to acknowledge my German *Doktorvater* again here. Miroslav Volf read the manuscript carefully and made major suggestions for revision, the wisdom of which is best indicated by the extent to which I have followed them. Kevin Vanhoozer proved that friendship need not conflict with the role of Series Editor. His voice has accompanied the entire writing process, and I am grateful for his guidance.

Discussions with one person in particular have influenced this book. Since 1987 Steven Knapp and I have systematically worked through most of the issues covered here, exchanging thousands of pages of correspondence on divine agency and related topics, single-handedly keeping our long-distance telephone company in the black. I hope readers will find traces of Knapp's well-known rigour in these pages; they will certainly find a few Knappian formulations sprinkled here and there.

This book is dedicated to my wife, Kate, who will find more of herself in these pages than in the other ones, though she has supported them all.

NOTE

1. See Arthur Peacocke, *Theology for a Scientific Age: Being and Becoming – Natural, Divine, and Human*, enlarged edition (Minneapolis: Fortress, 1993).

1

SYSTEMATIC THEOLOGY
AND POSTMODERNISM

THE POSTMODERN SHIFT

Once upon a time theology faced the 'scandal of particularity'. In New Testament times the scandal of particularity meant that the Greeks speculated about general principles of the universe (what could be broader than Aristotle's speculation on 'the theory of being *qua* being'?) whereas Christians preached Christ crucified. As Paul put it, 'Jews demand signs and Greeks seek wisdom, but we preach Christ crucified, a stumbling block to Jews and folly to Gentiles' (1 Cor. 1: 22f.).

The scandal was back in the modern period, after a hiatus of several centuries in which Christian thought had dominated in the West without seeming scandalous – for was it not obvious to everyone that God exists and that he is revealed only in the history of the Christian religion (and represented only by Rome)? Our early modern predecessors broke with the obviousness of the Middle Ages. For them it was *again* the general principles of reason that caused the scandal. Descartes, known as the father of modern philosophy, may have launched – in the popular mind, anyway – a tradition of reflection based only on the individual human subject and her quest for certainty. Yet *more* basic to his programme than the *cogito* was, as I have shown elsewhere,[1] the project of *ontotheology*: the attempt to derive as much knowledge of God

1

as possible from human reason alone. This tradition dominated modern thought in the West for 150 crucial years, and much more of modern philosophy owes its existence, its questions and its fundamental intuitions to this line of thought than is usually acknowledged.

Ontotheology seems to have met its match at the hands of Kant. Kant was the one who (again, according to the Received View at least) destroyed the ontological proof of the existence of God, the one who strew antinomies where once the evidences of the hand of God had been. Kant notwithstanding, most of the nineteenth century was still motivated by the perceived need to overcome the *pre-Kantian* modern tradition. Ontotheology represented for that tradition the highest claims of human reason: knowledge of the infinitely perfect being (*infinita perfectio*) who exists out of his own necessity; conclusive proofs of this being's existence; human nature as based on (and mirroring) this being; and all knowledge of the world as having its source in him. No wonder, in the face of such ambitious claims made on behalf of human reason, that Christian identifications of God as Trinity and of Jesus as the way to (knowledge of) the Father would represent a scandal for the Enlightenment!

But now the scandal is past – so, at any rate, my thesis. Why? because the lofty aspirations of metaphysical reason are now *passé*. The claims of reason to universal validity and its claim to be able to derive important truths in an a priori fashion, 'from reason alone', are now under severe challenge. Actually, the preceding sentence might be taken as a thumbnail sketch of what is widely called 'postmodernism', that apparently dominant movement that is receiving so much attention in Western culture today. Diversity is now the name of the game; multiple perspectives are the bottom line; every person and cultural group has his, her or its particular perspective; and we rejoice – at least in theory – in the differences between them all. There is no longer a scandal of particularity in a postmodern age because *it is precisely particularity that is being celebrated by postmodern thought.*

However, theologians have not yet caught onto this change of climate, it appears. By and large, they continue to march forward under the umbrella of *protecting* themselves and their traditions against the universal demands of reason, apparently not seeing that the sun is actually shining on particularity. (Liberal theologians seem to have recognised the new context, less so conversative theologians, who are generally those who aim to 'conserve' as much as possible of the historical tradition of Christianity.) Most traditional theologians continue to write as though their readers will find it scandalous that they begin with the biblical documents, documents with a highly specific history, a series of culturally influenced ethical claims, a saviour who lived at one particular time and place, a set of highly specific claims about God.

This defensiveness is a grave mistake. Precisely what postmodern readers *can* accept about Christianity (as also about Judaism and other religious traditions) is the fact that it has its own particular set of sacred scriptures and its specific cultural location. For, they will retort, does not *every* religion in the world have the same specificity? Postmodernism or 'multiculturalism', seen as a world-view,[2] accepts and rejoices in such specificity. To name one example: the American Academy of Religion recently approved a request to establish a major new 'programme unit' in Systematic Theology for the first time in several decades. The most important argument, it appears, was that Christianity should have its own particular tradition of reflection, its own theology, just as Judaism or Buddhism should and do within the Academy. Similarly, the study of Christian thought and practice can be found at an increasing number of public universities in the US – something that would have been unthinkable a few decades earlier – because from a multi-ethnic perspective there is no reason *not* to include this study alongside the others. Finally, Bill Moyers could present on National Public Radio in the US (beginning October 1996) a ten-part series on the book of Genesis, one which talked without embarrassment about the central stories and beliefs contained in the biblical texts.[3] The only requirements were that the participants represent a wide variety of religions and interpretive approaches to Genesis, that they not dismiss one another's views – even while they vehemently defended their own – and that they should treat the material as a series of stories rather than as truth claims that excluded all other truth claims.

Thus I suggest – to a degree that would have been unpredictable to most of us only fifteen years earlier – the doors have opened again for Christian theology. The long battle for Christian particularity, the context that forced Karl Barth to his famous 'Nein!' some seventy years ago, is over. Christian theology has *not* lost out to the universalising and homogenising forces of philosophy and natural theology, as theologians throughout the modern era feared it would.

Unsurprisingly, the meaning and implications of the term *postmodernism* are hotly debated today – as one might expect of the label for a broad (all-encompassing?) cultural movement. For example, Nancey Murphy distinguishes vehemently between 'Anglo-American' postmodernism and Continental (mostly French) postmodernism.[4] As a result, I suspect (following Wenzel van Huyssteen) that the term *postfoundationalism* would be significantly less misleading than *postmodernism*.

Van Huyssteen's *Essays on Postfoundationalist Theology* provide a clear portait of a theology that proceeds in critical dialogue with the sciences and philosophy – yet without the sense that theology's truth has to be derived at the outset from more general foundations.[5]

Whatever the label, the shift has unfortunately caught most theologians by surprise. They continue to fight for exactly what is now being offered to them free of charge. If we believe that old-style *Fundamentaltheologie* – prolegomena understood as apologetics – is no longer credible in the current context, we must begin in a very different way, naming our particularity right at the outset and without apology. What follows, then, is an extended reflection on the very particular Hebrew and Greek texts out of which Christianity arose, on a specific man, Jesus, around whose person and teaching Christianity revolves, and on the symbols and concepts that have created and informed the almost two millennia of Christian thought.

CAN THEOLOGY STILL MAKE UNIVERSAL CLAIMS?

I wish in no way to deprecate this new development called postmodernism nor to undervalue its significance. Theology now faces opportunities it has not known for centuries. But along with these new opportunities comes a new task: to rethink, and to re-present, what it *means* to make religious truth claims. For there *is* one particular move that is stigmatised in the contemporary climate: to claim that one's particularity has universal significance.[6] Christian thinkers have tended in recent years to move in one of two directions: either to formulate their truth claims in a more and more vehement fashion, refusing to give any ground and 'damn the consequences', belittling 'secular humanists' and 'the secular mindset' and anything that would challenge the tradition (the fundamentalist orientation), *or* to cease to make any specifically Christian truth claims at all, proclaiming themselves satisfied to represent one particularity alongside the others (the liberal response). In *A Far Glory* Peter Berger has given a brilliant account of these two tendencies and of the necessity of finding some sort *tertium quid* between them.[7] The very dominance of these two tendencies, and the fact that they are (obviously) dichotomised and hence share little or no common interests or vocabulary, has led to a certain scarcity of work on the project of finding middle ground. Interesting philosophical and theological resources exist for making progress, but too little work has been carried out in this area so far. Although this book will not address the methodological discussions as its primary focus,[8] it will be clear to the perceptive reader that these issues are not far below the surface. Better, it seems, to address the methodological issues by making some sort of progress on the questions themselves; to paraphrase the scholastic saying, the best proof of the possibility of something is actually doing it.

There is a related task, however, which *is* basic to the present project. Let us grant that Christianity does in fact make some truth claims, that is, claims which can in principle conflict with truth claims made in other contexts. The task then becomes not only to relate Christianity to other religious traditions, but also to the (in the public's mind, and perhaps rightly) major source of knowledge about the world: science. For, at the same time that cultural and religious particularity have received new credence, the authority of scientific conclusions as an overarching framework of knowledge has never been so great. This is no coincidence, of course; it is yet one more manifestation of the old fact/value distinction that emerges in the earliest texts of modern thought.[9] According to this view, there can be multiple ethnic and religious perspectives, and one can glory in this fact, precisely *because* they are all merely systems of values, myths, stories, interpretations. Because if none of these stories and none of their claims represent facts about the world, it is said, hence there is no reason that one cannot be a pluralist in all matters ethical and religious. The pluralism of cultures is not disturbing, one continues, because we *do* at least have a common understanding of how the world works: that provided by science. In short: one account of the world, many interpretations of that world; one physical world, many 'meaninged' worlds. If Christian theology is going to resist the widespread application of the fact/value dichotomy, it will have to carry the discussion beyond a 'self-explication of our particularity', and beyond inter-religious dialogue as well, to the level of a *fundamental discussion of the nature, status and truth claims of both theology and science.*

WHAT IT MEANS TO TAKE SCIENCE SERIOUSLY

The appeal to postmodern sensibilities – or to faith! – must not be allowed to make theology's task today easier than it really is. Though not necessarily antithetical to religion, the scientific mindset often stands in sharp contrast to the religious mindset. Pick up any issue of a popular science journal such as *Scientific American* or *Physics Today* and thumb through its pages. The articles are about empirical phenomena observable in the physical universe around us. Most of the articles describe a concrete discovery that researchers have made about this world or a pattern emerging out of a group of such discoveries. Each discovery is described in rigorous terms; it is reconstructed with the help of a theoretical framework that explains large segments of the empirical world; and it includes a full accounting of the physical states or events that caused the observed phenomena. Even when the proposed explanation is a speculative one – indeed, especially then – the author takes pains to show the conditions for checking her explanatory hypothesis empirically. She inevitably presupposes that explanations are strongest when they account for the widest possible range of empirical data with greater adequacy than any of their competitors.

Of course, a scientist is a person too. She has prejudices, she may hold religious beliefs, and she may wish or hope that certain results will turn out to be true and others false. Sometimes she may even be less than fully objective in her judgements about whether or not new evidence counts against or in favour of her own hypothesis. Yet when she is in the lab, it is her business to pay close attention to the data, to the viable explanations, and to the theories and hypotheses that best predict observed phenomena. If her theory predicts or explains less well than those of her competitors, the scientific establishment will eventually leave it behind. And if she falsifies her data, makes a computational mistake or puts forward empirical claims that are untrue, the odds are that her mistake will eventually be uncovered.

Scientists' concern with empirical data does not rule out any interest in broader explanations of the universe and humanity's place in it – explanations that may be, say, philosophical or religious in nature. Indeed, in so far as the book you are holding is located primarily within the discipline of theology, you have a right to expect that it will look carefully at a variety of explanations that go beyond what can be inferred, strictly speaking, from empirical data. Our goal will be to examine the doctrines of God, of God's relation to the world and of God's activity in the world. I will be presenting multiple positions, criticising some of them and offering a constructive systematic theology of my own.

But how is this possible, one might well ask. How can one take scientific results with the utmost seriousness, as I suggest we must, and still engage in constructive theological inquiry? The question is a serious one; it will occupy us in detail in the pages that follow. Imagine that one is convinced (as I am) of the importance of taking empirical inquiry seriously *and* of the importance of reflecting on broader theological and metaphysical questions. Even then, it is crucial not to deceive oneself about the sorts of answers that are most congenial to the scientist *qua* scientist – lest one pretend to satisfy science when actually ignoring it. Let us take one example: the study of 'mind'. Surely the phenomenon of thought – the individual experience of intentions, desires, ideas, volitions and the like – is more widespread than the experience of God (at least as classically understood). Yet note what the scientist *qua* scientist would most hope to discover about mind. As a scientist she would be most successful, by all accounts, if she were able to define 'mind' exhaustively in terms of empirically observable phenomena that admitted of clear patterns and were fully quantifiable. She would be even more pleased *qua* scientist if she could then find a full explanation of this clearly defined set of phenomena, where by 'full explanation' she would mean one that predicted them in terms of observable changes in the physical world (e.g. changes in brain states).

The goal I have just described is often called 'explanatory reductionism' or 'methodological naturalism'. Of course, the success of the reductionist program would not be good news for ordinary or 'folk psychology' – nor for theology, at least as the discipline has been traditionally understood! But there is no doubt that it would be good news for science as such. Bad news for science, by contrast, would be the discovery that mental phenomena are irreducible, that they obey (say) supernatural laws of their own, that their explanatory principles are inaccessible in principle to empirical study, or that they seem to be expressions of some kind of thing (say, the soul) that lies entirely outside of the physical world altogether. Clearly, there is some tension between the mindset of science and the goals of theology, at least at this methodological level.

Much of this book is about reasons to think that explanatory reductionism is, ultimately at least, inadequate. I will defend the theses that metaphysical and theological issues are raised by the methods and conclusions of the sciences; that theology and science share some basic principles of rational inquiry in common; that physical determinism and explanatory reductionism have (so far) been unsuccessful, and perhaps that they *must* be so in principle; that the world around us evidences sets of emergent properties that we call 'life' and 'mind' which exercise a causality of their own; and that the entire complex of data and experiences available to us are ultimately better explained in theological language than in the language of materialism and physicalism alone. Nonetheless, one must not make the task sound easier than the data actually allow. Theology may of course be pursued in abstraction from the results and methods of the sciences. But if we do choose to attempt a theology of nature 'in light of contemporary science', then we must not pretend the task is more simple than it really is. As we will find, the task requires an openness to scientific results and to the various directions in which they point; it necessitates that one wrestle with tendencies that run counter to traditional theological answers; it demands an openness to revising certain dearly held theological conclusions; and, at the points at which one may wish to break with the (apparent) implications of the scientific results, it mandates that one either find reasons *inherent within the sciences themselves* for making that break, or that one supply reasons that might be held to be convincing in other fields (history, the human sciences, ethics or philosophy) which point in the direction of the theological conclusions one wishes to defend.

The bottom line is that theologians must be intellectually honest with the data and conflicts that their discipline faces when it wishes to take contemporary science into account. It may be that the study of biological evolution is compatible with the notion of an overall purpose behind the universe; but it does not follow that the data of evolution *prove or give evidence for* the existence of such a purpose. It may be that Big Bang cosmology is compatible with the Christian doctrine of creation; still, the two are not identical. It may be, finally, that we are able to show that life is an emergent property out of physical matter, or even that consciousness is a real emergent in sufficiently complex biological organisms such as the higher primates; but such results would at best only suggest but would not prove that a spiritual dimension arises within consciousness at a particular stage. Modern physics *may* be more like Taoism than was Newtonian physics, and theism may be in better shape today than it was in 1690 or 1860. But modern physics is not *equivalent* to Taoism, nor does it demonstrate the truth of theism. Given that these sorts of conceptual and methodological differences exist between the sciences and theology, constructing equations must be seen as a category mistake. Science remains science and theology, theology; the compatibilities or implications one may be able to draw from one to the other do not eliminate the tensions that remain.

A word should be said about the human sciences. As I attempted to show in an earlier work,[10] the human sciences represent a crucial *tertium quid* between the 'hard' empirical sciences and theology or metaphysics. Science (in the sense of the Latin and German words for science, *scientia* and *Wissenschaft*, which mean the organised study of a specific domain of data) requires that one take the human being and its behaviours and actions as a specific object of study and devote disciplines such as psychology, sociology and anthropology to it. But if 'organised study' were a *sufficient* condition for a science, then T. F. Torrance would have been right to claim that theology is as fully scientific as physics.[11] What such easy equations do not mention are the specific strengths that characterise the empirical or 'hard' sciences – lawlike ('nomological') explanations, prediction, quantification, rigorous falsifications – which are not shared by any other area of organised inquiry. Social scientific explanations that play well in Western culture (say, Freudian psychoanalysis) may seem absurd to non-Western men and women, and what passes as a compelling metaphysical position in one culture may not seem at all compelling in another. Yet the explanatory power of, say, quantum mechanics is equally as strong in *all* cultures, and the leading figures of science are drawn from throughout the world. The fact is that the natural sciences have the tools to overcome differences of opinion in a way that is not matched by any other set of human disciplines. This is why I maintain that the debate with the natural sciences has a particular urgency for those who make theological truth claims, and I will pay particular attention to that debate in the pages that follow.

GOD ACTIVE IN THE WORLD

In the following chapters I will argue that two of the most urgent problems raised for theology by modern science are the problems of how to conceive of God's relation to the world and how, *if at all*, to conceive of God's agency in the world. The former is a well-worn theological theme and thus needs no introduction, but the latter deserves a word here. Gone, it seems, from the popular mind, at least in the developed West, is the belief that God is constantly involved in changing the outcome of natural processes in the world. By contrast, for the 'pre-modern' mind, for which 'natural laws' were merely summations of how human experience habitually turns out, and for which the human and divine realms so thoroughly permeated one another that no firm dividing lines could be drawn, a constant involvement of a divine agent or agents in human experience had seemed unproblematic.

But those days are past – even for many believers who would like still to save a place for the notion of divine action in the world today. The reason is not simply that our knowledge of natural processes and their regularity has increased, though this is certainly the case. Nor is it due only to the theological argument that a God who found it necessary constantly to meddle in the order which he had created would be inferior to or less perfect than a God who got it right the first time about. Just as much, the change has been brought about by the *mindset* underlying modern science and technology. It seems that the expectation of the regularity of nature, the belief that objects act according to pre-dictable laws, went a long way toward making possible the scientific study of nature (and the concomitant implementation of technology) in the first place. There would be no point in your repeating an experiment over and over if you did not have the expectation of discovering what causes the regularity and why the observed outcome occurs in precisely this way. Equally, there would be no point in attempting to reconstruct events of the past, to understand the causes that led to particular human behaviours, if you did not believe that there *were* causes which lend themselves to subsequent historical-critical reconstruction. It is impor-tant to note that the mindset I am describing does not try to rule out divine action on metaphysical grounds; it is not a form of metaphysical naturalism. Instead, it is a way of thinking or methodology deeply ingrained in our culture, educational system and habits of thought – a set of expectations that makes science as we know it possible at the same time that it makes traditional accounts of divine action problematic.

One more reason for the difficulty of claims about divine action must be mentioned: the problem of evil. Whenever a theologian claims a large amount of divine involvement in the world, she will find herself con-fronted in an urgent way with the question of God's *non*-involvement on particular occasions. If God is never active in the world, then no explanation needs to be given of his non-involvement at a particular moment. If God is active on only the rarest of occasions – say, only at crucial moments in salvation history – then one can understand why he would not be involved in preventing, for example, a fatal traffic accident for a young family. God's silence is just par for the course, one must respond. (Of course, this is not to say that deism does not face other problems of its own.) But if God is *constantly* altering natural events to bring about goals of his own, and if he often answers prayers – arranging the weather for the benefit of my family vacation, healing my cold so that I can speak at the conference, miraculously repairing the church heating system so that no further financial investment on our part is required – then the question becomes all the more urgent: why is God *not* active in a similarly visible way when the bomb is directed toward the shelter full of orphans? Why would he not end the life of the dictator whose orders are responsible for the deaths of millions?

Theologians have produced a multitude of answers to the problem of evil, some more convincing than others. The first standard to be applied to such answers is consistency: theologians and believers must think of God as an agent *whose actions are consistent with the nature that one attributes to him*. If one offers a 'rich' account of divine action, according to which

God is frequently altering states of affairs, then one must also be prepared to provide an account of why he does not act at other times. The form of the problem of evil based on divine *inactivity* is the most significant criterion for control over the entire discussion of divine agency. It should be said in advance: where I am hesitant to claim that 'God acts in this particular way' in what follows, a major source of caution remains the fear of making God responsible for cases of apparently random suffering. This theological concern remains in the background even when I am citing scientific difficulties. The suffering of innocents should constantly accompany the theologian as she writes, and it is far better to be silent than to be glib in one's responses to it.

To many theologians, and of course to many more outside of the scope of theism altogether, the problems seem insurmountable. Frank Dilley gives particularly powerful expression to the dilemma:

> it becomes increasingly obvious that the alternatives are only two: either a conservative tradition affirming miraculous acts of God, whether spectacular or 'hidden', or a God who acts solely through the general orders and processes of nature and history. The former alternative runs counter to the anti-miracle tradition in modern thought, to our unwillingness and inability to cash in miracles, as well as counter to our theological distrust of a God who is a being beside other beings. If the times can be reversed and men can think miraculously once again, then there is still place for a God Who Acts. If he cannot, then a God who is conceived as a general cosmic process (perhaps personal), who works universally rather than specially, is the only hope. Such a God might produce unique currents of history through variability of human response to him, but not through variability of his own action.[12]

The really insightful essays into the problem of divine agency over the last years wrestle with these two alternatives and with the forces that would compel theology to the one or the other answer. Together with some other recent authors and in dialogue with them, I shall be searching in these pages for a position that can do justice to *both* – one that neither abandons the agency of God altogether nor holds onto the full knowability of miracles in the traditional sense. Is it enough if the world is such that we can *view it as if* God were causing events? Would it suffice for God to effect *psychological* miracles – unexpected hope and faith in the individual mind – without acting in any other way within the world? Could God direct universal history 'from the top down', as it were, without any concrete interventions in that universe? Or does

11

Christian theology require believers to assert that God acts on a regular basis as a direct causal agent within the physical world? It is as important for theologians to grapple with the real difficulties facing the doctrine of divine agency – theological, conceptual, scientific – as it is for them to formulate constructive responses to the challenges.

NOTES

1. See Philip Clayton, *Das Gottesproblem: Gott und Unendlichkeit in der neuzeitlichen Philosophie* (Paderborn: Ferdinand Schöningh Verlag, 1996); in English: *Infinite and Perfect? The Problem of God in Modern Philosophy* (forthcoming).

2. See Charles Taylor, *Multiculturalism: Examining the Politics of Recognition*, ed. Amy Gutmann (Princeton, NJ: Princeton University Press, 1994).

3. The series has since been published by Bill Moyers as *Genesis: A Living Conversation* (New York: Doubleday, 1996).

4. See Nancey Murphy, *Anglo-American Postmodernity: Philosophical Perspectives on Science, Religion, and Ethics* (Boulder, CO: Westview Press, 1997). Whatever our disagreements on names, Murphy provides a very clear account of the epistemic situation for theology in the present.

5. See Wenzel van Huyssteen, *Essays on Postfoundationalist Theology* (Grand Rapids, MI: Eerdmans, 1997).

6. The matter is actually somewhat more complex, since certain moral claims, including the claim that it is wrong to deny anyone her particularity, *are* in fact made and condoned by postmodern thinkers. This fact leads to some ironic situations, such as the situation of American scholars condemning women in tribal cultures for being closed to the sorts of roles that women play in American culture today.

7. See Peter Berger, *A Far Glory: The Quest for Faith in an Age of Credulity* (New York: Maxwell Macmillan International, 1992).

8. I have attempted that task in some detail in *Explanation from Physics to Theology: An Essay in Rationality and Religion* (New Haven, CT: Yale University Press, 1989).

9. See especially Louis Dupré, *Passage to Modernity: An Essay in the Hermeneutics of Nature and Culture* (New Haven, CT: Yale University Press, 1993). The current view receives masterful expression in John Searle, *The Construction of Social Reality* (New York: Free Press, 1995).

10. Clayton, *Explanation*, Chapter 3.

11. Thomas F. Torrance, *Theological Science* (Oxford: Oxford University Press, 1969).

12. Frank B. Dilley, 'Does the "God who acts" really act?' in Owen C. Thomas (ed.), *God's Activity in the World: The Contemporary Problem* (Chico, CA: Scholars Press, 1983), pp. 45–60, quote p. 58.

PART I

*The God Who Acts:
Towards a Biblical Theology of
God and the World*

2

WHAT IS THIS WORLD
WHICH THOU HAST MADE?
GOD'S RELATION TO THE WORLD
IN THE HEBREW BIBLE

—————

> In the beginning God created the heavens and the earth. The earth
> was without form and void, and darkness was upon the face of the
> deep; and the Spirit of God was moving over the face of the
> waters. (Gen. 1: 1f.)

God created the heavens and the earth. In this statement we find perhaps
the clearest expression of the Lordship of God, the absolute dominance
and primordiality of the divine source. But we also encounter an area in
which a final dichotomy between truths of reason and truths of faith,
between science and religion, would make no sense. Of course there
may be competing *emphases* and areas of specialisation in the two
approaches, but there can be only one physical world. This world either
had its ultimate source in natural processes of immense physical energy,
mass and time, or it had its source in some sort of intention or purpose
– say, in the conscious decision of a personal being.

Thus we cannot accept a dichotomy of the sort propounded by the
theologian Paul Zimmerman:

> It would seem better to accept in simple faith the account of the
> origin of man as given in the book of Genesis, to allow scientists
> to continue to pursue their research, but to insist on what Scripture

tells us concerning man's origin and his purpose in life as well as the redemption which God has wrought for us in Jesus Christ. It would be tragic if we were to permit scientific theories, scientific philosophies, to dictate our theology either explicitly or implicitly. Let us really let the Scriptures speak to us and not attempt to read into them many things which even liberal commentaries insist the authors never thought of.[1]

A much better model of the necessary connection and interplay between scientific and religious accounts of the world is offered by Wolfhart Pannenberg. Pannenberg warns that theology

> should . . . not fall into the all too comfortable escape, the mistake of developing a theology of creation on a particular and exclusively theological level which is inaccessible to all criticism by the natural sciences, for instance as a commentary on the first two chapters of the Bible . . . The confession of the God of the Christian proclamation as the Creator of heaven and earth remains empty, remains a mere confession of the lips, as long as we do not have good grounds for asserting that the nature with which the natural sciences are concerned indeed has something to do with this God.[2]

This sense of the universal significance of creation stands in opposition to the approach to the doctrine of creation taken by Karl Barth, who consistently understands creation as the 'external ground or framework for the covenant'.[3] Barth correctly understands creation as a part of the history of redemption, which allows him to tell a consistent story of the salvific purposes of God in history. But the stress on redeeming covenant leads Barth to pass too quickly over the world and its reality as an independent theological locus, even to the point that he sometimes refers to creation as '*pre*history'. By contrast, it will be our attempt here, without denying the many relations between the two doctrines, to resist the subordination of creation to the doctrines of redemption, salvation and sanctification. Theologians must grapple with the explanatory task posed by the existence of the world, with its rich patterns of mathematical predictability within the physical phenomena, its brutal struggles for survival in the biosphere, and the complex and confusing behaviours of the human beings who dominate this planet. If we do not work hand in hand with those who have rigorously studied nature in all its guises, listening to these specialists at least as much as we ask to be listened to, then we have no chance whatsoever of developing an adequate theology of nature.

THE DOCTRINE OF CREATION IN THE HEBREW BIBLE

Of course, there will be no dialogue if the theologian has nothing to say. Only if we have some sense of what the Christian tradition has proclaimed about nature, its source and meaning, will we have a starting point for a productive conversation with the natural sciences of our day. The task of this chapter, then, must be to consider the teachings on the creation of the world, and the place of humanity within it, that are found in the creation narratives in the Hebrew Bible or what Christians have called the Old Testament.

The Doctrine of God

The first thing to note is that God's role as Creator was not separated from his covenantal and redemptive role in relation to humanity. As Anderson writes in *Creation versus Chaos*, 'it is curious that in Isreal's faith during its formative and creative period (1300–1000 BC) the belief in Yahweh as creator apparently had a secondary place' to a belief in Yahweh as historical saviour.[4] Two conclusions immediately follow: first, that whatever doctrine of the Creator we develop will have to be closely linked to the other roles that God plays in the Hebrew scriptures; and second, that the textual accounts of Yahweh must not be read as dispassionate quasi- (or pseudo-)scientific accounts of the world. Surely this is what is right about Barth's stress on the notion of covenant: Israel knew her God first as a saving, guiding and protecting presence, and only later, gradually and by implication, as the Creator of all that is.

The Israelite tradition had reached strict monotheism by the time the oral traditions were recorded. We find in the Hebrew scriptures only small remnants of the polytheistic religious background that dominated the Near East during this period, such as in the use of the plural *Elohim* in Psalm 82: 1 ('God takes his stand in his own congregation; He judges in the midst of the gods').[5] What characterises God's relationship to the world above all else is his *Lordship*: because he is known to his people as Lord, God must also be the source of the world and everything in it. He does not share his power with other gods; he is limited neither by them nor by any human being or human circumstance. As Robert Butterworth summarises the Old Testament picture, 'There is one God alone who totally transcends all created reality and has uncontested dominion over it. There is no question of a time when he did not exist: he is pre-existent, equivalently eternal. His creative action is spontaneous and unassisted by anything outside himself.'[6]

At the same time, with a few exceptions to be considered below, monotheism in the Hebrew Bible was not developed in a philosophical direction. One finds no speculation on whether God is absolute perfection, whether he is a being or Being Itself, or which attributes should be predicated of a being who is both infinite and perfect. At the same time there are of course numerous passages in which the Israelite writers have reflected on the nature of God and what follows from it. In particular, they show a clear awareness of the implications of God's absolute transcendence of the world. For example, nothing in the world is to be worshipped *as God*, and no inner-worldly object can convey the divine glory or truly express his nature (Exod. 20: 4).

Moreover, nothing in the world can *constrain* the action of God. Everything he carries out has its ground in his own free decision. Thus Claus Westermann writes, 'The transcendence of God over all space and time implies that the interaction of God with the universe he has made rests on the free ground of his own eternal being.'[7] It would never occur to the Israelite writers, as it did to the Greeks, to assert that the intractability of matter made God unable to make the world any better than he did or unable to carry out his will in it if he should so wish. Transcendence over the universe implies God's unlimited power. If we must recognise any limitation, it must be, as Torrance has emphasised, a limitation on the part of the universe.

Reconsidering this last point brings us to perhaps the most important implication of God's unlimited power for a doctrine of nature: the contingence of the world. Everything that exists, exists because of the divine concurrence; without the continued Yes of God, none of creation would remain. Pannenberg has thus made the notion of contingence the central feature in his doctrine of creation.

> [W]e must presuppose the contingency of all world occurrences and not just individual events. This presupposition is what we find in the biblical belief in God as the Creator of the world. This belief finds in the incalculability and contingency of each individual event an expression of the freedom of the Creator. Hence with Schleiermacher, and with Augustine before him, we will find the fact of the order of nature, its regularities and enduring constructs, genuinely astounding . . . With the contingency of occurrence, then, a direct relation of all individual events to the divine origin of all things is posited, notwithstanding the part played by creaturely secondary causes in what takes place.[8]

18

Its contingency, in short, more than anything else, marks the created world, and this contingency pertains to every feature of the world. When physicists emphasise the contingent nature of the world,[9] they employ a concept that has deeply theological roots. As Torrance emphasises:

> The understanding of the contingent nature of the cosmos, upon which all empirico-theoretical inquiry rests, derives not from natural science but from Judaeo-Christian theology, i.e. from the doctrine of God as Creator of the orderly universe, who brought it into existence out of nothing and who continuously preserves it from lapsing back into chaos and nothingness.[10]

Theological anthropology can be conceived of as a detailed inquiry into the implications of the contingency that is basic to human nature. As Westermann notes, 'However, just as order in the universe is contingent, so the *freedom* in the universe is contingent and therefore limited.'[11] It is a contingent fact of this contingent world that we have been created free. According to the tradition, God has chosen to limit his power in order to leave room for the actions of created creatures. Some events God simply allows to occur on their own – not because he lacks power over the wind and the waves, but because humans can only be free in a world in which there is room for them to use their own powers of moral choice and action. Moreover, there would be no place for human agents in a world that God ruled by caprice as a total tyrant. Only if some events occur in a lawlike manner, only if there is a predictable context of regularity, could humans have developed the expectation of being able to carry out actions on their own. We should therefore endorse the conclusion that 'Considerations such as these on the role of "chance" in creation impel us to recognize more emphatically than ever before the constraints which God has imposed on himself in creation and to assert that God has a "self-limited" omnipotence and omniscience.'[12]

How Did Creation Occur, and Out of What?

The theology of creation is about the grace of God, about God's love and positive plans for humanity. Reflection on creation arose for the Israelite people out of their experience of God's covenantal relationship with them. In many ways the belief in creation seems to have been an extrapolation backwards from what were their experiences in the midst of history. If Yahweh was powerful enough to create all things whatso-ever, then he must also have sufficient strength to tend his people; if Yahweh demonstrates unmatched and apparently unlimited power in guiding his people, then he must also have been the originary power that called *all things* into existence in the beginning. Von Rad reflects the logic of this argument: 'Yahweh the Creator, who raised up the world out of chaos, does not leave Jerusalem in chaos.'[13] The doctrine of creation, therefore, represented a salvific promise in the *Sitz im Leben* of the Israelite people.

Creation is also about the *sovereignty* of God. As Y. Kaufmann notes, 'The basic idea of Israelite religion is that God is supreme over all. There is no realm above or beside him to limit his absolute sovereignty. He is utterly distinct from, and other than, the world; he is subject to no laws, no compulsions, or powers that transcend him. He is, in short, non-mythological.'[14] In many early creation accounts or 'cosmogonies' running from the Mesopotamian and Egyptian to the Greek, the creating god or gods are taken to be limited by a pre-existing 'stuff' of some kind or another. By contrast, the Israelite tradition came eventually to be characterised by its view that the act of creation was absolutely unconstrained. One sees this, for example, in the use of the Hebrew verb *bara*, which refers to a work of God that is unique, carried out without effort on his part and relying on no matter or conditions outside himself. For Scheffczyk *bara* implies 'the unique sovereignty of God over what he has made, a power limited by no antagonistic primordial principle.'[15]

The Judeo-Christian tradition thus breaks from many other ancient cosmogenies in its strict insistence on *creatio ex nihilo*, creation out of nothing in the strictest sense. This does not mean, writes Torrance, 'that [God] created the universe out of some stuff called "nothing," but that what he created was not created out of anything'; the existence of the universe was entirely dependent upon God's will to create it, and the universe as a whole is therefore a contingent entity.[16] Moltmann adds, 'The later theological interpretation as *creatio ex nihilo* is therefore unquestionably an apt paraphrase of what the Bible means by "creation." Whatever and when God creates is without preconditions. There is no external necessity which occasions his creativity, and no inner compulsion which could determine it.'[17]

The doctrine of *creatio ex nihilo* is not itself taught in Genesis. Now Gen. 1: 1 – 'In the beginning God created the heavens and the earth'– is often taken to imply a creation out of nothing. But the text does not directly require this interpretation, and it is in fact not until much later writings (e.g. 2 Macc. 7: 28) that the doctrine of a creation out of nothing is explicitly stated. Instead, it appears to have been the *cumulative picture of God* as it emerged in the Hebrew scriptures and as it is reflected in the opening chapters of Genesis that led to the inference that God is so in control of history that he could not be limited by any other principle. Thus Scheffczyk concludes, 'The term *bara* so forcefully conveys the Creator's independence of any other creative principle that this text (Gen. 1: 1) has usually been taken to imply a *creatio ex nihilo*.'[18]

This very indirectness of the answer to the question, 'Out of what was the universe created?' is significant for our reading of Genesis. It has often been said that Genesis was never written (or told) as an argument in natural science or historiography, at least in the modern, historical-critical sense of the word. That *creatio ex nihilo* – and the divine attributes that follow from it – are never directly expressed in the biblical creation texts is a clear reminder that they are not philosophical texts either. Genesis is best understood as a look 'backwards' to the beginning of time and the world by a people who knew themselves to be called out as a 'people of the covenant'. Whether one holds, as many conservatives do, that God revealed the story of creation directly to the biblical writer(s) or, with more liberal thinkers, that the creation stories represent an extrapolation backwards out of the Israelites' later experience of Yahweh,[19] one can grant that Genesis is first and foremost a *story of God's creative and redeeming acts*, not an objective account of the universe's origin but a highly reflected and 'mythologised' set of conclusions about the divine purposes behind the universe. As Gilkey wrote, '*creatio ex nihilo* seemed to many intelligent Christians, as well as to secularists generally, to be one of those early mythological notions which the Israelites expressed with their usual poetic and religious sensitivity.'[20] Creation out of nothing by an infinite and perfect being is the *post factum* conclusion, not the starting point, in the Hebrew account.

The theological significance of a creation out of nothing cannot be overemphasised. If God created strictly out of nothing, then *everything* has its source in him – not only natural laws, living beings and all objects, but also space and time themselves. There is no primordial matter to set limits on God or to resist his creative wishes, and *everything* that exists, the objects as well as the laws that govern them, owes its existence solely to God. Thus Torrance rightly emphasises, 'The creation of the universe out of nothing involves the creation of space and time as well, which means that they are to be regarded as orderly features of empirical processes or events within the universe and not as detached empty "containers." '[21]

Continuous Creation

The mediating moment between the original instant of creation and the subsequent doctrine of providence – the belief that God continues to guide and protect his people through history – lies in the doctrine of continuous creation (*creatio continua*). The theologian cannot help but notice that the acts of creation ascribed to God in the Hebrew Bible, both in Genesis and in later passages such as the creation psalms (especially Ps. 8), involve long-term plans and intentions. These divine intentions could not have been completed in the original six 'days'; they would have to be worked out over a broad span of human and cosmic history. 'Hence new realities come into being, and old ones often pass away, so that God's action as Creator is both past and present – it is continuous.'[22]

So continuous creation reflects, in part, the acknowledgement that the process of forming the world as God wished it, described in Genesis 1, could not have been completed by the time of the creation of humanity. The same God who was responsible for setting the various cosmic forces in motion would have to continue to guide them and would have to produce new things as long as this world order continues to exist. Yet the doctrine of *creatio continua* is much more than a logical implication of Genesis 1–2 or a nod in the direction of what we now know to be scientific fact. It acknowledges as well that one phase of God's relation to the world (initial creation) is past and that another one (his providential care for what he has made) has begun. At the same time it recognizes that some features of divine intention and action – namely, God's nature and his creative activity in the world – embrace and run through both phases. Scheffczyk spells out these relations clearly:

> This last addition [God's Sabbath rest after creation] has often been interpreted too negatively, as if it simply announced the end of the act of creation in temporal history. But God's sabbath rest has a thoroughly positive meaning: it declares not that the world is unfinished, but that its basic structure is complete and that creative activity has now been extended so as to establish a new relationship between God and the world.[23]

The reason this link is theologically significant is that – in the Israelite context, as well as for the dynamics of faith – the actions of God in creation *must* be adequate to tell us something about what we have to expect from God in the ongoing process of history. The trinitarian principle employed by Karl Rahner ('Rahner's Rule') suggests that we must be able to make judgements about the immanent Trinity from the economic Trinity, inferring to (towards) God's ownmost nature from the actions that God carries out in the world. Analogously, we must be able to draw inferences from God's acts in creation to his providential care for the created world. As Gabriel Daly notes, 'When one sees Creation in close connection with *preservation*, then one forestalls a rather obvious distortion of the biblical Creation statement, namely that the reflection on Creator-Creation is merely giving information about the origin of the world and of man.'[24] The doctrine of creation is never a matter of abstract knowledge of the world; whatever claims it may make about how the physical universe came to be, it remains first a *religious* claim, an assertion that involves the core being of the subject who asserts it. For the Israelites, the repeated statement that frames the creation narrative in Genesis 1, 'and God saw that it was good' (Gen. 1: 4, 10, 12, 18, etc.) thus implied the promise of continuing care for that very creation. One sees this same pattern of thinking in the Psalms: the creative activity of God is mentioned in the context of praise for the providential care of God. Thus the creation psalm begins and ends with, 'O Yahweh, our Lord, How majestic is Thy name in all the earth!' (Ps. 8: 1, 9), because creation is taken as a promise of God's continuing availability and care. This is the religious appropriation of the creation statements stressed by Luther (among others): the same God who worked the mighty acts of creation is *pro nobis*, continually present with us through the Spirit.

What characterises the doctrine of continuous creation is that, more clearly than the initial act of creation, it connotes that God uses indirect means to bring about creative purposes within the world. Why is this different from the claim that God created the universe in the beginning? Physicists call the initial moment at which the physical universe comes into existence '$t = 0$',[25] since there is no meaningful way to speak of physical time before this point. Before $t = 0$ physical space and time do not exist and no physical laws apply, so that physical theory has no purchase here. Even the claim that 'something' caused the Big Bang to occur is not a well-formed physical statement. In so far as the causality referred to precedes any physical conditions that could determine it, it must pass as a metaphysical rather than a physical statement. At the beginning, if anywhere, is a natural point to speak of the direct creative activity of God, who produces the conditions for the entire physical universe out of himself with no mediation or dependence on outside forces.[26] (Indeed, the proximity of Big Bang cosmology to theistic creation is so great that some physicists have rejected it for this reason alone. The astronomer Hoyle, for example, made no secret of the fact that his alternative 'Steady State' model was developed in order to avoid opening the door so clearly to theological claims.)

But what of the creative actions *subsequent* to that initial act of creation? Here, I suggest, we should follow Moltmann's interpretation of continuous creation, namely as a type of divine creation that is not contingent upon God's *direct* action, but upon the processes established in the parts of nature that are already created. Moltmann writes:

> If we are trying to find a new interpretation of the Christian doctrine of creation in light of the knowledge of nature made accessible to us by evolutionary theories, we must distinguish more clearly than did the traditional doctrine of creation between creation in the beginning, continuous creation, and the consummation of creation in the kingdom of glory.[27]

There is no theological difficulty in acknowledging the indirectness of continuous creation; the theological question rather concerns whether the *results* of the process can be attributed to God's plan for and providential care of creation. *How* God achieves these results is a scientific question. (The same applies, incidentally, to the manner in which the biblical texts were inspired and came to be authoritative for the Christian tradition.) One can imagine the reactions of Habakkuk's original Israelite hearers to the proclamation that 'I am raising up the Chaldeans, That fierce and impetuous people Who march throughout the earth To seize dwelling places which are not theirs' (Hab. 1: 6). That God has used the instrumentality of humans to achieve his plans is basic to the biblical understanding of providence. Why then challenge whether he could use natural laws and physical processes – either before or after the advent of humanity – to achieve the divine plan?

Which is More Fundamental, Creation or Redemption?

It is not true to say that Israel's religious faith was from the outset a faith in God as the Creator of all. Hebrew Bible scholars have long agreed on the subordinate status of the creation theme in the Hebrew documents. G. von Rad's formulation is classic:

> Our main thesis was that in genuinely Yahwistic belief the doctrine of creation never attained to the stature of a relevant, independent doctrine. We found it invariably related, and indeed subordinated, to soteriological considerations ... [B]ecause of the exclusive commitment of Israel's faith to historical salvation, the doctrine of creation was never able to attain to independent existence in its own right.[28]

When one looks more closely, most of the Hebrew texts about creation are actually working to convey something about the human condition, the God/humanity relation or God's redemptive aims and acts toward his people.

The biblical writers began with the context in which they lived and through which they knew God: the covenental relation with Yahweh. Their reflection was fuelled by the memory of key events in their own tribal history. In the midst of history, surrounded by hostile tribes, threatened by their own tendencies toward syncretism, they attempted, and often failed, to keep the exclusive covenant with their God. Their reflection on the origin of all things arose not out of the attempt to gain mastery of the world through science nor out of an objective (historical-critical) interest in 'what actually happened' at some point in the past, but rather as the effort, *ex post facto*, to comprehend who this God was who demanded such obedience. One might say that the perspective in the Hebrew Bible is the opposite of the problems that have dominated the modern world. Whereas the intellectual challenge of modernity has been deism – the fear that, although God may have created the world, he has not done anything in it since then – the inference within the Hebrew Bible moves in the opposite direction: the Israelites gradually conclude that the one who redeemed and rules them was also the Creator of all that exists. In this sense Butterworth writes: 'It was not in the wonders or in the order of nature that Israel first came to know God, but in his saving actions in history on Israel's behalf.'[29] In an important sense, *creation recapitulates redemption*. More exactly, the doctrine of creation recapitulates the doctrine of redemption; it re-expresses the experience of redemption and the inferences based upon that experience in the context of the cosmological question, 'And where did everything come from in the first place?'

This difference of emphasis allows for the Israelite writers to use creation metaphors freely when speaking of the creation of the people of Israel. In fact, it is almost true to say that creation metaphors have their *primary* reference here, their secondary reference in the creation stories of Genesis 1–2. For in one sense the Exodus was the time of Isreal's creation *ex nihilo*, the time when God created a people out of nothing.[30] A similar concept emerges in Paul: 'And it shall be that in the place where it was said to them, "You are not My people," there they shall be called sons of the living God' (Rom. 9: 26). Similarly Simkins concludes, 'By using creation metaphors to express these subjects, the biblical authors have presented the human condition and redemption *in terms of* God's activity in creation ... Creation in the Bible therefore serves as a paradigm or model of the human condition and of redemption.'[31]

Perhaps we could distinguish between the way that Israel developed the doctrine of creation and the use to which it was later put (think of it as the difference between the context of discovery and the context of justification). The faith in creation arose out of cultic practice; it was a 'doxological' belief, one intrinsically reflecting the context of worship.[32] To slightly (but only slightly) overstate it, the creation doctrine arose as a *present-tense doctrine*, one less about an initial act of God than about the ongoing dependence of the nation of Israel, and the individuals within it, on their Creator. Consequently, it was *as a redeemed people* that Israel came to reflect upon the origin of the world and to link its cultic God with the origin of all that is. Indeed, it is fair to say that this extension was an intrinsic part of the development of absolute monotheism in the Hebrew Bible.[33] Note that this movement is matched, at least approximately, in the New Testament, where the redemptive work of Jesus in Mark is extended past Matthew and Luke to the principle of the *logos* as described by John and by Paul to the idea of the pre-existent Christ (e.g. Colossians). Of course, in both cases the movement is not merely linear: once the notion of God the Creator has been developed, it can be followed outward, in a conceptual movement of its own, into an entire systematic theology. Is this not what Westermann implies when he writes, 'So the activity of the God of Israel is extended to an activity in the history of the nations and beyond to Creation, preservation, and blessing throughout the universe . . . When the Psalmist sings . . . [that] this praise is extended to the lord of the history of the nations and to the Creator of everything created, *it is but the same process* . . .'[34]

It seems then that a theology of divine action must include two theological movements: a *movement of recovery* backward from redemption to cosmology, from Redeemer to Creator; and a *movement of providential care* forward from creation to God's trustworthiness within history. We also find in the Hebrew Bible, however, a careful concern to keep nature, as one source of knowledge about God, subordinated to God's direct dealings with Israel – as one finds also in Paul (say, in the opening chapters of Romans) a distinction of 'that which is known about God . . . since the creation of the world' (Rom. 1: 19f.) through nature from the revelation in Jesus Christ. Why the concern with the limits of what can be known through nature? Von Rad suggests, '. . . the doctrine of redemption had first to be fully safeguarded, in order that the doctrine that nature, too, is a means of divine self-revelation might not encroach upon or distort the doctrine of redemption, but rather broaden and enrich it.'[35] The texts reflect a fear that created things rather than the Creator would be worshipped, a fear spawned by the availability of animistic religions and reflected in the commandments against graven images: 'You shall not make for yourself an idol, or any likeness of what is in heaven above or on the earth beneath or in the water under the earth. You shall not worship them or serve them' (Exod. 20: 4f.). Paul is equally vehement about the danger of 'exchanging' or confusing creation and Creator:

> Since the creation of the world his invisible attributes, his eternal power and divine nature, have been clearly seen, being understood through what has been made, so that they are without excuse. For even though they knew God, they did not honor him as God . . . Professing to be wise, they became fools, and exchanged the glory of the incorruptible God for an image in the form of corruptible man and of birds and four-footed animals and crawling creatures. (Rom. 1: 20–3).

This fear is an important source of the Israelite, and later the Christian, tendency to construe God as completely separate from the world. It is a motivation that we will want to look at very closely in later chapters – especially if we find that the conceptual context that made that fear necessary has changed, or that new options are available for thinking about God that neither identify God with his creation nor separate him completely from it.

THE WORLD'S RELATION TO GOD

Is there a difference between God's initial creation and his subsequent intervention in the world; and if so, how should it be conceived? Can God's actions during creation be characterised as somehow different from later interactions with his people? Is the creation fully separate from him, or is it in some sense 'within' him?

The Israelite doctrine of creation offers a double message. On the one hand, Israelite cosmology stood out boldly against its neighbouring cosmologies for its refusal to identify God with any part of the world whatsoever. Arguably, as the first among the world's religious traditions Israel proclaimed the complete transcendence of God over against the world. Later – at the hands of the more philosophically minded Greeks – this emphasis hardened into philosophical dualism, the doctrine that a cleft exists between the finite world and its true source. The later Christian doctrine of 'creation out of nothing' was developed against the backdrop of this Greek dualism and came to be understood as an expression of it. Thus, reflecting backwards through this history, Gilkey interprets *creatio ex nihilo* as the means for saying that God is *not* immanent in the world: 'Understandably, therefore, many liberal religious leaders objected to the "dualism" of the idea of *creatio ex nihilo*. For in that impassable gulf from the temporal, finite world He creates.'[36]

On the other hand, *only a being like Yahweh who is not identified with (and in this sense reduced to) any particular object in the world could be present in all objects of the world.* This is the doctrine of divine omnipresence:

> Where can I go from thy Spirit?
> Or where can I flee from thy presence?
> If I ascend to heaven, thou art there;
> If I make my bed in Sheol, behold, thou art there.
> If I take the wings of the dawn,
> If I dwell in the remotest part of the sea,
> Even there thy hand will lead me,
> And thy right hand will lay hold of me.
>
> (Psalm 139: 7–10)

Such a doctrine could only arise in the context of a religious belief like Israel's, a belief that refused to identify God's essence with any object or event in the world. (Likewise, it could easily be shown that the same logic applies to the other 'omni's' as applies to divine omnipresence: God's omnipotence, omniscience, omni-benevolence, etc. presuppose this first moment of difference from the world.) Thus the very separation that theologians speak of as the transcendence of God from the world, the separation that gave rise to *creatio ex nihilo*, also makes possible, finally, a closer connection between God and world than had been the case for Israel's neighbours:

> The God who made the world and all things in it, since he is Lord of heaven and earth, does not dwell in temples made with hands; neither is he served by human hands, as though he needed anything, since he himself gives to all [persons] life and breath and all things. [Further,] he made from one every nation of humankind to live on all the face of the earth, having determined their appointed times and the boundaries of their habitation, that they should seek God, if perhaps they might grope for him and find him – though he is not far from each one of us. For in him we live and move and have our being, as even some of your own poets have said, 'For we also are his offspring.' Being then the offspring of God, we ought not to think that the divine nature is like gold or silver or stone, an image formed by the art and thought of man. (Acts 17: 24–9)

In the passage – Luke's recounting of Paul's so-called Areopagus address in Athens – we see the progression clearly: God is not to be identified with any object, holy place or human projection. Rather, he is the source of all things that exist, and their *telos*. But for this very reason, he is also not to be separated from them; instead, he is so close that he is within all things, and only in him (and because of him) do they live and move and have their being.

Moltmann is perhaps the most forceful recent proponent of the doctrine of the indwelling of God within his creation:

> By the title, 'God in Creation,' I mean God the Holy Spirit. God is 'the lover of life' and his Spirit is *in* all created beings . . . This doctrine of creation, that is to say, takes as its starting point the indwelling divine Spirit of creation . . . The Creator, through his Spirit, *dwells in* his creation as a whole, and in every individual created being, by virtue of his Spirit holding them together and keeping them in life.[37]

Here we see precisely the closer relationship developed in the preceding paragraphs. The God who is not confused with any object in the world – the One who is infinite where they are finite and necessary where they are contingent – is the one who can and must sustain all individual things. We see in the biblical texts, in short, the gradual emergence of an ontological dependence of *all* things on God. This doctrine has the structure that we have observed throughout this chapter: Israel, called out of nothing and utterly reliant on Yahweh for its survival against overwhelming odds, gradually began to conceive the entire world as equally dependent on an analogous concurrence, a continuing positive willing of creation by God. Just as Abraham, Isaac and Jacob owed their very existence to God, who made them into a people and protected them from their enemies, and just as the people followed Moses into the wilderness, receiving not only guidance but also food (Exod. 16) and clean water (Exod. 15) from God, so also, the biblical writers inferred, the whole world must rest on the grace and the concurring acts of God. Think of it as a sort of theological 'anthropic principle': the world itself is inhabitable, made for humankind and good, *because* of its reliance on God and the presence of his Spirit or breath within it.

The Israelite authors bring us to this point but they do not take us beyond it. God is both transcendent and immanent, they teach, and yet the implications of this both/and are not fully spelled out by the authors – perhaps the question is too speculative for them. It might seem to us, for example, that a God whose Spirit is so closely identified with creation needs to be affected, and even altered, by what happens in it. It might also seem a natural development to think of God as suffering along with his creation, as has often been observed: 'An immanent Creator cannot but be regarded as creating through such a process *and so as suffering in, with and under it*.'[38] Nonetheless, the idea of God becoming focally incarnate in the world, even in the form of a particular person, or the idea of modelling God's relation with the world on an inner dialectical (trinitarian) structure, lies beyond the Hebrew texts. So too do arguments about the God/world relationship drawn from natural theology and science, as in the claim, 'Thus it is that the scientific perspective obliges us to take more seriously and concretely than hitherto in theology the notion of the immanence of God as Creator – that God is the immanent Creator creating in and through the processes of the natural order.'[39] Still, the seeds of such further reflection have been planted. It has been the task of later theologians to water them, and some of the fruits of this ongoing conversation may only be ready for harvest in our day.

HUMANITY AND THE IMAGE OF GOD

We have already seen that the book of Genesis is not to be read as a science textbook. Its lessons are theological; it is not in competition with natural scientific accounts of the origin of the world. (This is not to say that there *could not* be conflicts between the natural scientific account and the theological principles – but more on this later.) When Genesis states, '"Let there be a firmament in the midst of the waters, and let it separate the waters from the waters." And God made the firmament and separated the waters which were under the firmament from the waters which were above the firmament' (vv. 6–7), this clearly represents a now outdated picture of the nature of space and the relationship between the oceans of earth and the heavens. Nor should we be over-quick to determine which aspects of this account remain relevant and which not. Surely theologians still want to say that God is responsible for the creation of the heavens as well as the earth, but do we also need to insist on a distinction between the nature of the one and the other? Or should this text be used to defend a fundamental ontological connection between the two? Since the physical world-view on which this text is based is one that we no longer hold, it would seem arbitrary to claim to possess an authoritative manual of translation that could pick out precisely which elements remain authoritative and which not. The text calls for a theological reading, yet to say this does not yet determine the degree of detail that can be theologically appropriated nor the closeness of the theological reading to the text's originally intended meaning.

The descriptions of the emergence of life (11f., 20ff.) present a beautiful picture of a planet swarming with life, of species multiplying and filling every available ecological niche. Is the *order* in which these living things are introduced theologically significant? Are they understood to form a sort of hierarchy of emergence, or is it merely the picture of the richness of this divinely orchestrated symphony that is meant to catch our attention? The text provides little help here. Thus we must be careful to avoid two related errors: on the one hand, to declare that a text that is scientifically inadequate can have nothing theologically significant to say about the emergence of life; or, on the other hand, to proclaim that a particular theological reading of the significance of the individual 'days' of creation and the steps in the process have a normative weight to the exclusion of other possible readings. The text can neither be tied to a particular scientific hypothesis nor be cast free from all significance that might lie in its detail. Unfortunately, no a priori means lie at our disposal for establishing the exact balance between these two errors.

What then is the *theological* significance of this account of creation? The following points represent some of the more important implications:

1. In one sense the act of creation represents the absolute transcendence, and even contradiction, of what it is to be human. To be human is to exist as conditioned, as always already thrown into the midst of a world, a culture, a historical epoch, a family, a socioeconomic status, a race. Even before one makes one's first choice, one is conditioned in almost all aspects of one's being. The Creator, by contrast, is constrained by none of these things. God's nature alone dictates what he will create; his absolutely free and unconstrained choice by itself determines whether he will create at all. The theological/philosophical terms 'finite' and 'infinite' could well be introduced and defined by this very distinction between created and Creator.

2. At the same time, God's act of creation presents a powerful model of who it is that we are and what the image is in which we have been created. Bonhoeffer writes:

   In the beginning God created heaven and earth. That is, the Creator creates – in freedom! – the creation. The connection

31

between the two is conditioned by nothing other than by freedom, i.e., it is unconditioned . . . Between Creator and creation
is absolutely the Nothing. For freedom plays its role in and out
of the Nothing. Thus we can find no necessities within God,
which could, or even would *have to* lead to the creation; there is
simply nothing that grounds the creation.[40]

Creation takes place not through a giving birth or a vomiting up
of the world, as in other creation myths of that time. Instead, it
takes place by the Word of God – a thought in the mind of God,
intended as action by God, spoken by God as order and as sign of
his will. God's act of creation becomes the quintessential sign and
symbol of the free, rational, ethical agent who forms goals and
carries out actions.

3. The act of creation is a sign of unlimited power. Certainly, as we
have seen, this idea is expressed by the doctrine of *creatio ex nihilo*.
But it is equally implied by the other strand in the Genesis creation
story: the creation out of darkness, out of chaos, *tehom, tihamat* –
ideas that have their sources in the Babylonian 'primordial ocean'.
In either case God starts with what is antithetical to God – absolute
Nothingness, or what is dark, empty, chaotic – and produces a world
in his image. This act already introduces an ethical dimension, the
ability to transform what is opposed to God into something in his
image and useful to him. It is not too much to say that already in
the act of creation we witness a *redemptive* moment, an anticipation
of the later redemption of the world in Christ. The doctrine of
creation (and of nature) is not yet the doctrine of salvation and
redemption; nonetheless, creation is itself, in this broader sense,
already redemptive.

4. Creation is in and of itself an ethical act, because its Creator is
Goodness itself and because nature is declared good from the very
outset (vv. 5, 9, 11ff.). From the start what God has produced is
declared good. The physical world is not for the biblical writers a
matter of an ethically neutral 'stuff'. Hence, although we need not
say that from the moment of creation the fall was predetermined
(superlapsarianism), we do need to acknowledge that the possibility
of a fall was programmed into creation from the start.

But it is not only against value-neutrality that we must beware;
it is also against the tendency, recurrent throughout the history of
Christian reflection, to see the physical world as somehow evil, or
at least inferior to the realm of pure spirit. The Platonic roots of

this tendency are well known: Plato spoke of the physical world as illusion, as inferior and as only partially real in comparison with the realm of the forms, which alone represented the locus of all value, truth and reality. The Christian and Jewish doctrine of nature stands diametrically opposed to this tendency. God saw it – the physical world itself, the various levels of life as it emerged, and the man and woman who most approximated God's image within nature – and proclaimed them good. Their *telos* was not to be something they were not, was not to leave behind the physical realm as inherently evil and strive for a disembodied state; it was instead to be most fully in the image of God *as physical/spiritual entities*. A rash of heresies can be avoided when one pays sufficient attention to the doctrine of creation: soul was not created to be separate from body and to rise upward to a realm of its own; universal history does not have the goal of leaving the physical universe behind, but rather the goal of returning it to the perfect state in which it was created; and God does not view the physical universe that he created as separate from himself and thus evil, but rather as itself good – and, in this sense at least, a part of himself. 'Because the world is God's, it is good. God intends a good world, a good work – he who is creator and lord of the world. The flight from the created work into a disembodied spirit, into a pure meditation, is forbidden. God wishes to behold his work, to love, preserve, and call it good.'[41]

The Place of Humanity in Creation

It is significant that in Gen. 1: 26ff., for the first time, the author of the narrative reports God as stating his intentions: 'Let us make man in our image . . .'. It is the God who acts not as a blind force, not as a deterministic (or random) law that gives rise to each stage of random evolution, but as a free, consciously choosing personal force – it is this God who creates humans in his image.

To grasp the *imago dei* one must grasp that the continuity of humans with the rest of creation is not unlimited; at one point at least an important *discontinuity* emerges. This is the point where humans are understood theologically as made uniquely in the image of God, as set apart in some way from the rest of creation. The early Christian tradition tried to express this difference by using the categories of Greek philosophy: Plato's realm of the forms as the natural home of the soul, and Aristotle's notion of man as 'rational animal', set apart from nature by his possession of *ratio*. Neither of these captured the essence of a Christian theology of nature.

Yet Genesis strongly singles out the creation of man and woman from the rest of creation. This is the one species which is made in the image of God (Gen. 1: 26f.), the one created to 'have dominion' and to 'rule', for whom all else is food (1: 29), the one that calls forth for the first time the use of the intensifier 'very good' (1: 31), the one who names the animals (2: 19f.) – and the only one immediately shackled with a limitation, a law to obey (2: 17). The inference lies close at hand: 'The human being is given his high status not for his own sake, but in view of his divine commission to be master over other living creatures.'[42] One is struck by the immediacy of the God/humanity relationship; it is as if the writer is saying to his hearers:

> Look, the peoples around you find their gods in animals and hills and heavenly bodies, in the seasons and the forces of nature. You must be utterly different from them. For you not only must nature be de-divinised; it must not even be the occasion for your coming to know God. You must meet your God face to face and without mediation.

If this characteristic of the Judeo-Christian tradition opens theology up for reflection on the divine nature apart from the world, as we have seen, it also contains the seeds of an alienation of our species from the rest of nature. Recall the famous passage from Harvey Cox's *The Secular City*:

> When he looks up to the hills, Hebrew man turns from them and asks where he can gain strength. The answer is, Not from the hills, but from Yahweh, who *made* heaven and earth. For the Bible, neither man nor God is defined by his relationship to nature. This not only frees both of them for history, it also makes nature itself available for man's use.[43]

We have direct access to God, and the relationship to God is the most important aspect of our existence; hence is it not of little consequence how we treat the natural world around us? Thus it was that the Judeo-Christian tradition came to be blamed as the major cause of the environmental crisis. Now it may well be, as I also think, that this is a misreading of Christianity, and that the tradition contains theological resources for an environmental ethic as strong as that of any tradition.[44] But it should also be admitted that these are *indirect* implications of the

Hebrew texts given that the primary message about humanity in Genesis is the direct, unmediated relationship with God.

Not that humanity is unlimited! We have already seen that the very first thing God says to the human in the Gen. 2 account is a commandment. For the man the enjoyment of creation comes with strings attached. This is a threefold limitation: it limits the man's actions, what he can do in this new world; it threatens him with judgement, with an *imposed* limitation like that of a parent on a child, should he transgress his boundaries; and it defines him *ontologically* as a contingent being limited by death: 'for in the day that you eat from it you shall surely die' (2: 17). Westermann comments insightfully, 'Man, just because he has been created, carries within him limitation by death as an essential element of the human state, important for the course of the narrative. The Flood narrative also indicates quite clearly that the history of mankind will have an end.'[45] Ours is the species placed under, and defined by, the onus of mortality.

Scholars have long agreed on the diverging emphases of the two creation narratives. Genesis 2.4b–24 is significantly older and places humanity much more at the centre of the account; Genesis 1.1–2.4a is later and emphasises the cosmic framework much more strongly.[46] It would, however, be a mistake to conceive of the Genesis 2 account as 'anthropocentric' and Genesis 1 as focusing on God alone in abstraction from God's relation to humanity, as some scholars have done. Both sides of this contrast need to be looked at more closely: the place of humans in Genesis 2 is clearly found in humanity's subordination to God, just as recent discussion of Genesis 1 has brought out its anthropocentric perspective in a way that was not as clear in traditional accounts.[47] The struggle to pull the creation accounts in one direction or another is painfully clear in the recent literature. Don Cupitt, for example, sees in this account the first inklings of 'the production of reality by language', the process of making God in the image of man, which has come to fruition in our own time: 'Like us,' he claims, 'God is made only of words. So . . . a new sort of theology and a new doctrine of creation begin to emerge. For we can no longer distinguish clearly between the sense in which God creates, the sense in which language does, and the sense in which we do.'[48] Far from the account of God as human projection given by Cupitt, Labuschagne challenges the reading of even Genesis 2 as anthropocentric: 'One of the most deplorable misconceptions with regard to the biblical doctrine of creation is that creation is usually considered to be anthropocentric . . . Far from being anthropocentric, the biblical view of creation is *theocentric*.'[49]

It may just be possible today to specify a moderate position that expresses the motivations for *both* of these positions. In an appropriately limited sense, both 'God creates humanity' and 'Humanity creates God' have their element of truth. God has created the universe and is responsible for the eventual emergence of all life forms, including humanity – this is the basic assertion of theism in all its forms. If God is not in some way more than the universe as a whole, *and* more than any human picture or theory of God, then theism is false. The flip side of this belief is the realisation that our theories of God are, at least in part, *our* models – conceptual constructions to which the theologian herself contributes. It is no more true to claim that some 'pictures' or concepts of God are free from all extrapolation out of human experience than it is true to say that all ideas of God stand on the same level because all are *nothing but* human projections onto empty heavens. Genesis is about a free, rational creature whose function it was to name 'every beast of the field' (2: 19f.) and whose failing it was to wish 'to be like God, knowing good and evil' (3: 5). But above all it is about the One who was before any finite things were and who chose to create all that is, including – highest among them all, and yet often the lowest – that one species that was called 'in Our image, according to Our likeness' (1: 26).

We need to struggle with what it means to read Genesis today from a standpoint beyond the old tug-of-war between anthropocentrism and theocentrism. Part of this new reading is to hear in a new way what it means for Genesis to proclaim that God was before man and woman, before the animals, before all things. Such a reading will 'demote' humanity to its rightful place as one among countless species on whose interdependence the entire biosphere depends. And yet the demotion of humanity is not so severe as to remove the eschatological promise that echoes through the biblical documents. For these texts simultaneously offer the promise that, long after all things have crunched together into a point of infinite mass and temperature, or (under another cosmology) settled into an unending state of nearly motionless bits of matter a few degrees above absolute zero, God will continue to preserve whatever is most essential to the identity of those life forms that lived for (at most) a few billion years in the middle phase of the universe's lengthy history.

Personhood as the Image of God

We have examined the nature of humanity in light of the concept of the 'image of God'. The significance of this issue for understanding the relationship between God and world cannot be overestimated; the *imago dei* question contains the kernel for all later thought about the God/world relationship. In no version of Christian theology, no matter how revisionist, are God and world identified. God's divinity, based as it is on an infinite nature that always stands above and beyond the finite nature of contingent beings, requires theologians to draw a clear distinction between the divine and the human. At the same time, the image of God connotes the continuing presence of God within his creation, the refusal fully to separate Creator and creature.

According to Genesis, God is present to all of creation. Yet 'God-likeness', even when it applies to all of creation, applies most distinctively to humanity. We have already noted a number of features of humanity that reflect the divine nature: humanity's moral nature, its rationality, self-consciousness, responsibility to others and to the earth – and its freedom. Of all of these, it is perhaps freedom that most succinctly expresses that unique state of being which is being a person in the image of God. As Bonhoeffer writes, 'That God creates his image on earth in humanity means that humans are similar to their creator inso-far as they are free.'[50] The author of Genesis describes only one being who, waking up to find itself in the midst of Paradise, is immediately confronted with a decision: to obey or not to obey, to accept its created status or to seize the knowledge of good and evil that belongs to God alone. Freedom is the leitmotiv of theological anthropology, the theory of personhood: we are free to worship God; we are free to make rational and moral decisions; and we are free to turn away from God, to alter the image that was created within us.[51]

Note that personhood is intrinsically a *relational* state: one is not a person in and of oneself, in abstraction from relationships; rather, one is a person from and for one's interactions with other persons.[52] It is a kind of relationship that is constituted simultaneously by freedom and responsibility. In Bonhoeffer's words, 'freedom is in the language of the Bible not something that humans have for themselves, but something that they have for others' (p. 41). Ultimately, we are free *for* a particular type of relationship with God. And we are free *not* to enter into that relationship – as paradoxical as it may be, we are free not to be that which we are and were created to be. Herein lies the 'moment of truth' in the thought of existentialists such as Jean-Paul Sartre: being a person is not something that can be defined out of a timeless essence (though here the theologian would add: at least not by this means alone). Rather, personhood means to *choose* who one is through one's relations with others – and, one might add, with one's God. It means to have a self-conception and to attempt to act in a manner consistent with that self-conception and its values.[53]

What sets humans apart from the other products of creation, then, is not a disembodied soul, a hunger for a purely spiritual state of existence, or an ontological gap between us and the so-called lower life forms. Instead, it is a particular feature of the image of God that is not shared (at least not in the same sense) by other life forms, a difference that we point to with the label *personhood*. There is a common nature shared between God and the animals, but humans alone are the ones who are addressed by the divine *word*. Humans are creatures who are able to stand outside of their existence in the world, to be conscious of who made them, of what the Creator's (expressed) purposes were, and what it is that they are called to do in the world. Anthropologists speak of *homo sapiens* as the species least controlled by instincts; we are the animal whose behaviour, to a remarkably high degree, must be learned anew through enculturation and training.[54] Philosophers have glossed person-hood in terms of the qualities of *freedom, self-consciousness,* and *moral responsibility*; theologians have supplemented these categories with the uniquely religious component of the *imago dei*, the capacity to love (God and others) as God as loved. The *positive* side of this human uniqueness is that we are able to take on a special place of responsibility within the meta-ecosystem that is the earth, to care not only for others of our own kind but also for the planet as a whole and for its inhabitants. It is when we carry out these responsibilities in a God-like manner that we are 'little less than the angels' (Psalm 8) and the *imago dei* is visible in us. The *negative* side, however, is that it also lies within our power, as a consequence of the very same freedom, that we can act in opposition to our created status and the nature that was given to us.

Positive and negative – the story of creation is thus simultaneously the story of the fall, potential and actual. Only one animal was created such that it could fall, namely that animal whose nature it is to constitute herself as person *vis-à-vis* other persons and in encounter with her God. Herein lies the sense in which humans are free and responsible in a way that other animals aren't – not that our behaviour is better (it certainly isn't!), but that we stand continually before the decision of how we shall act and constitute ourselves by that action. The implications of this unique responsibility, both in the ethical realm and in the realm of human reflection, will occupy us in later chapters.

THE FALL

The Universality of Fallenness

Theology is about redemption. Since the beginning it has sounded a positive note about the prospects for humanity and the world, thanks to the workings of grace and a sufficiently long period of time. And yet . . . humankind is fallen. The story of the fall plays a major role in the Christian account of the God/world relation, for it is simultaneously the story of *the continual falling away of humans from God*. This is the heart of any Christian theological anthropology: that the human creature diverges repeatedly, if not inherently, from the Creator. Here is sounded for the first time the theme of 'the destruction of creation at the hands of the created ones'.[55] Arguably, it was not until the advent of environmental science and a new understanding of the interconnectedness of all living (ecological) systems that the universal logic of this pronouncement was fully understood, namely: when humanity falls, all fall. The advent of technology and the reality of the human ability to take the entire planet with us in our selfish pursuit of comfort and personal gain is only the outworking-in-fact of what was proclaimed theologically in these very first texts of our tradition.

We thus turn to the theme of the fall. Consistent with the method outlined earlier, we will not preoccupy ourselves with the question of whether there was a historical fall. The primary significance of the doctrine does not lie in the affirmation that once, at the origin of the world, there was a paradisiacal state and then, at some later time, there was a fallen state. One may remain neutral on this particular question, though I am inclined to think that the knowledge of the evolutionary mechanisms (and the evolutionary struggle) basic to all life forms should now incline us away from the claim that there was a point in time at which all life dwelt in a state of paradise. At least, if there *was* once a primordial paradise, not only is there no evidence for it, but it also flies in the face of the bulk of what we have been able to ascertain about the nature of the biosphere.

Be that as it may, the fundamental theological point is clear: humans are morally and spiritually responsible before their Creator. Since moral responsibility presupposes some ability to choose, the concept of freedom plays an important role in the discussion of the fall. It stresses that one part of creation in particular, humanity, faced its Creator and his laws – and even its own nature – through the relation of choice. 'And the Lord God commanded the man, saying, "You may freely eat of every tree of the garden; but of the tree of the knowledge of good and evil you shall not eat". . .' (Gen. 2: 16f.). This creature was created free to accept the constraints its Creator placed upon it, to live obediently before God, and thus also free to turn away from God, altering the order within which it was created.

The theological traditions diverge on the question of whether the fall 'might not have happened.' A branch of the Reformed tradition (super-lapsarianism, discussed above) stresses that from the moment of creation the fall was predetermined; other theologians have written of the fall as a permanent aspect of creation, even at the risk of thereby making God responsible for it. I think it is possible to think of the fall structurally (rather than historically) without 'blaming' God for fallen humanity. Because humans are free, we are responsible for the actions that we take and for their consequences; indeed, we were *created* with this responsi-bility. Further, it is clear that we often (almost always?) choose to do something other than what would be the perfect response. We live continually in a fallen state. Here one is inclined to be more critical of Augustine's doctrine of original sin: no metaphysical entity or waiting period is required to explain the continual falling from moral perfection that characterises human action, for each human being provides more than enough ground for the doctrine in her own free decisions to do what is less than the best.

What is most significant about the doctrine of the fall is the element of sin. Sin connotes actions which (1) are freely chosen, (2) are the responsibility of the agent in question, (3) fall short of the moral ideal or what one ought to do; and (4) violate not only responsibilities that we have to each other and to the earth, but also our responsibilities toward our Creator. The fall represents a change in our status *vis-à-vis* God himself; it says that the moral imperfection and selfishness that characterise our actions have a significance that is more than merely human or inner-worldly. In part, this is an observation that transcends history, one that flows from the biblical claim that our moral nature is defined in contrast to a Being whose nature it is to be fully perfect. In part, however, 'fallenness' also has the more specific or historical meaning that humans have broken standards established (contingently) in the creation of the world – standards built into humanity by the very nature with which we were created. Again: historical fallenness does not require a historical fall; it is sufficient if the doctrine expresses a basic feature or characteristic of human beings in the world, confirmed anew in the experience of each individual. The key point theologically is that fallenness expresses a characteristic of humanity that is dependent on a 'vertical' comparison: the contrast with moral perfection, instantiated (according to the Christian tradition) in a morally perfect being. It is only the vertical dimension that justifies the use of the predicate 'evil', which involves not only the judgement that actions are damaging to others, but also the judgement that they are intrinsically bad, that they go against the fundamental moral order of the universe.

This interpretation of the doctrine of the fall may not make an original state of sinless existence and a subsequent historical fall essential to the theological narrative. Still, like the traditional view, it does insist on both a logical (or theological) 'moment', the inherent moral difference between God and humanity, *and* a historical moment, the individual experience of falling and fallenness. It may be that the particular narrative in Genesis 3 is to be read mythically, as a composite picture of humans turning away from God and seeking to be (in Augustine's terms) *causa sui*, the cause of themselves. But, *contra* Bultmann, the divine response cannot be interpreted in a purely existential sense. If God is active in any way in the world (if deism is false), then the answer to the human state that Christianity has called sinfulness must include a divine initiative aimed at the redemption of individuals and of creation as a whole.

We might put this point differently: given an (individual or communal) fall, and given the existence of God in anything like the sense that theology has traditionally maintained, one must retain a place for acts of redemption. Basic to the characterisation of God in the Jewish and Christian traditions is the belief that God is involved redemptively with fallen humanity, with a creation whose moral nature is often opposed to his own. In light of God's moral perfection and God's providential concern for creation, we cannot speak of the fall and a state of separateness from God without also imagining an overall narrative in which history is being drawn in the direction of a final overcoming of the barrier raised by sin. The opening chapters of Genesis tell the story of a God who from the very first moment of recognising humanity's fallen state is committed to acting salvifically to overcome the moral difference – and yet in a manner that respects human freedom. God's iniative then becomes grounds for action on the part of believers. In the New Testament, Paul famously turns the *indicative* of the fall into the *imperative* of the call to believers: 'Put off your old nature which belongs to your former manner of life and is corrupt through deceitful lusts, and be renewed in the spirit of your minds, and put on the new nature,

41

created after the likeness of God in true righteousness and holiness'
(Eph. 4: 22–24). Note how the *imago dei* here becomes part of a moral
exhortation, an exhortation to righteousness.

A Proclivity toward Evil?

The theological tradition has often speculated as to what part of
humanity is fallen and what part remains above the fall. Calvin stresses,
for example, the darkening of the intellect;[56] the Augustinian tradition
has stressed the fallenness of the will, in some cases arguing that it
inclines toward evil, in others that it has lost all ability to act in accor-
dance with the will of God (total depravity). The philosophical tradition
has gone both ways, running the gamut between humanistic visions of
humans' unlimited ability to improve their status on the one hand, and
ethical egoism and hedonism on the other – that is, positions that assume
the complete selfishness of humans on all points of motivation. In a theology developed in dialogue with the sciences, it becomes an *empirical*
question to what extent humans continue to act altruistically and to
what extent their actions are selfishly motivated. Thus biologists have
begun to specify the extent to which the goal of gene survival and
propagation is consistent with altruistic behaviour of particular individuals or phenotypes.[57] Greater attention to empirical results will save
theologians from the embarrassing situation of attempting to make
pronouncements about how humans actually behave based on the
theological tradition alone and in abstraction from the careful study of
human behaviour.

The study of history provides perhaps the most damning indictment
of humans' proclivity toward evil. In historiography we find an all too
clear record of what humans are capable of doing to one another and
what, over the long haul, the patterns of our interacting have been. It
has been said that the history of the twentieth century alone would be
sufficient to establish the fallen state of humanity *vis-à-vis* its God and
to instantiate most of the theological components of the doctrine of
the fall! Psychological, sociological and anthropological studies – from
comparative cultural studies of childhood aggressiveness through studies
of mob behaviour to statistics on rape and family murders in various
societies – add concrete evidence of human patterns of behaviour, supplying data for broader reflection on the nature of the human animal.[58]
Our greatest self-justifications and most noble moments are marred, it
appears, by a pervasive selfishness and dishonesty even to ourselves, a
self-deception that inclines one to respond, with the creed, 'and there is

no health within us'. In short, the social sciences arguably provide further support for a doctrine of the fall understood in the sense presented above.

In fact, it would be possible to show that each of the major *loci* of the traditional doctrine of the *imago dei* within humanity finds corresponding expression in the doctrine of the fall. We remain rational creatures, yet our rationality is swayed by personal interests; we remain self-conscious beings, though we manage to deceive ourselves about much of what is within us; we remain fundamentally moral beings, yet we are characterised in many respects by selfishness; we continue to have the call to work and cultivate the earth, yet we transform this call into an excuse for domination both in our relations with each other and with the earth itself. As in photography the image on the negative anticipates the final picture although everything is in reverse colour, so each of the 'fallen' features of humanity reflects the God-given capacities of our original image, yet in reverse. A remnant remains and is visible in each quality that humans manifest, individually and corporately – and yet it remains true that no part of the remnant is left untouched by the principle of sin.

The fundamental consequences of the fall are three:

1. Humans are left alone. Bonhoeffer writes, 'Man is *sicut deus*. Now he lives out of himself; now he creates himself his own life; now he is his own creator. Needing the Creator no longer, he becomes his own creator insofar as he creates his own life' (p. 90, my trans.).

2. Humans are set in opposition to their fellow humans. Life becomes the war of all against all, in which the principle of social interaction becomes the gain of the individual rather than the good of the whole. This enmity that enters into the core of human relations cannot be overestimated; it appears to be the guiding principle of the fallen state.

3. Humanity is alienated from itself, set at odds with its own nature. Augustine emphasised selfish desire or *concupiscentia*, humans' attempt to be the cause or principle of themselves (*causa sui*). Unfortunately – so the biblical message – to attempt to be the cause of oneself is to act in a manner inconsistent with our own nature as created. Thus we are attempting to be what we are not, to play the role of God when we are not gods. This self-alienation is the clearest expression of the fallen state, even if not its most insidious.

We have looked in some detail at the doctrine of the fall because of its centrality for understanding the God/human relationship. This relationship is seen most clearly now in its 'inverted' form. In other words, we can see most clearly how we are destined to be related to God by seeing how it is that, at present, we standardly live in opposition to that destiny. According to the theology of nature, the relation of finite creatures to an infinite Creator has an ideal structure – as does, for example, the correct behaviour of a finite being who remains dependent on a finite (and sensitive) environment and on limited natural resources. Yet what we find in human history, in politics and in the most basic interactions of humans with each other is evidence of a finite creature acting as if it were an infinite being, unlimited by anything other than its own will and self-affirmation (Nietzsche). The recognition of this state of being 'out of synch' is perhaps the most crucial Christian insight into human nature; it forms the core of theological anthropology.

The doctrine of the fall is thus central to Christian theology. It speaks a powerful word in an age that has grown tired of unlimited technological self-affirmation and suspicious of melioristic calls to the infinite improvability and trustworthiness of the human race. From this standpoint of the 'inverted' relationship with God and in light of the ideal expressed in the fiction of an original Paradise, we must now think our way toward an understanding of the correct relationship. A key task of the theology of nature is to formulate the ideals for the God/human relationship – ideals that we can hold up as models for future human thriving in both the horizontal and the vertical dimensions.

Is Genesis Historically Accurate?

We have not attempted in this chapter to draw that distinction which is foreign to the author of Genesis, the distinction between what is historical and what is 'merely' pictorial or metaphorical. Modern science has taught us that many things recorded in the opening chapters of Genesis did not take place in precisely the sense described in the texts. However, I presume the original authors and redactors would have been happy to concede this point. Their task, and thus our interpretation of what they have written, was in the first place *theological*; it was the story of the interactions between a redeeming God and that God's people. Consequently, we have asked what these chapters have to say about the nature of God and, in particular, about the God/world relationship. It should be too obvious to require saying, but 'theological' does not mean 'made as abstract as possible so that it has no discernable import for the reader'. When the theologian looks for the theological significance of the metaphors and narratives in Genesis, abstracting thereby from the question of the precise scientific truth of the matter, her aim is to *intensify* the existential involvement of the reader. God's purpose in creating the world and humans, the nature of nature, the image of God in humanity, the moral responsibility and moral shortcomings of humans, and the state of human existence *post peccatem*, after the fall – all of these topics are issues that impact on current readers with much of the same force with which they impacted on the original readers.

The story of the creation is also the story of the fall; creation and fall belong closely together. Yet the description of the fall – and every description of sinfulness – is in a particularly strong way the expression of an ideal, however shattered. What we take away from the opening eleven chapters of Genesis is not only a description of human beings who continually turn away from their Origin and Source, but also an implicit description of an (originary *and* final) 'paradise', an envisioned state in which God and creation are not separate. Because God is for the Israelite writers a redeeming God, every description of fallenness and sinfulness is inherently eschatological: it concerns a final state where the difference and the incompatibility of God and humanity will be overcome. *This* level of narrative – the one that stretches from creation to eschaton – and not the level of precise scientific explanation is the one that can truly pass as theological.

FIRST IMPLICATIONS FOR A THEOLOGY OF NATURE

We have seen that the doctrine of creation is an extraordinarily rich theological locus. It expresses what must be the first insight about God, the beginning of the Creed: 'Maker of heaven and earth' – that God is the absolute source of all that is, the power that preserves it in existence, the *ratio* or guiding principle that underlies and sustains it. A number of other classic loci follow almost directly as inferences from the theology of creation. If God is the source and destiny of all finite things, then he is also the source and destiny of humanity (theological anthropology); if he created the world with a certain purpose and goal, and he is all powerful, then he will also continue to sustain and guide it (the doctrine of divine providence); given these purposes and the immutability of the divine character, he can be trusted to bring his purposes to culmination at the end of history (the doctrine of eschatology).

The Nature of the World

We also know a number of things about what is in the world from our understanding of how and why, and by whom, it was created. These are, so to speak, the fruits of the doctrine of creation. We know that all finite things have their reason for existing, and their model for existing, in their divine source. We know that, as finite, they are distinguished from their infinite ground; although they are God-breathed, they are not identical with the full nature of God. They are, second, temporal: they have their origin at some concrete moment in the past. Third, they are eschatological or future-directed; they are defined by the process which will culminate, if God is God, in the eschatological state which is the goal of human and cosmic history. Fourth, they are not merely material (although they are that) but also spiritual; they are the handiwork of a being who is spirit. Fifth, they are moral beings: their Creator is Goodness itself who created them with a moral purpose. Thus they are themselves moral: defined by a moral ideal, by the proclivity towards sin, and by the centrality of the categories right and wrong within their very nature. Sixth, they have not been left alone. The One who created them has also promised to remain with them in this creation and to guide them. The One whose nature is love has not abandoned humans to their own powers (though he requires that they exercise them) but promises through providential care and guidance to enable them, to the extent that they are willing, to actualise the nature that is within them. Seventh, it follows that creation cannot be separated from the questions of salvation and the completion of history at the eschaton. God in his creation always knew that there was the possibility that these finite creatures would fail freely to conform themselves to his image, and in their failure they would need to rely upon his grace and the salvific work in Christ. Finally, the doctrine of creation has everything to do with practical theology: with ethics, political theology, social philosophy, environmental ethics, and the ideal for interpersonal relations.

The Divine/Human Relationship

In the doctrine of creation the full complexity of the God/humanity relationship comes to light. We have just noted that the creation is in its very nature distinct from its Creator; as finite it is distinct from the infinite. At the same time, 'creation out of nothing' implies that we are composed of nothing *other than* God; no pre-existent matter served as our building blocks. Of course, *creatio ex nihilo* does not automatically mean *creatio ex deo*; it does not mean that we consist of God. But it does

mean that we have no cause – no matter, no form, no goal and no efficient cause – save God and God alone. Everything of which we consist
is from God. This is not pantheism; but it *does* bring God significantly
closer, ontologically, to his creation than a 'Greek' notion of creation
with its assertion of the eternal existence of matter.

Indeed, the fear of pantheism led theologians for too long to separate
the world *too* severely from God, leading to an overly strong doctrine of
the transcendence of God that is incompatible with the fundamental
insights of a biblical theology of creation. (The influence of the Greek
doctrine of a pre-existing matter on the church fathers cannot be overestimated in this regard.) Only in the last few centuries has a full and
adequate awareness of the *dialectic between transcendence and immanence*
emerged. God is fully immanent in the world; in him we live and move
and have our being – and this in the strictest sense: to the extent that we
have being at all, we are 'composed' out of him who is Being itself. To
use a spatial metaphor, we are 'in' God, although he also extends beyond
us; to use a physical metaphor, we are like God's body, although he also
extends above and beyond what is this (finite) body. Increasingly, theologians are designating this position as *panentheism*, the view that the
world is *in* God, although God is also *more than* the world. We return to
the evolution of this doctrine in later chapters.

Interestingly, the same dialectic applies to beings in the world: they
are on the one hand completely dependent on God for their existence,
and on the other they are required to take their own actions (and to be
defined by them) as independently existing beings. The way in which
God has graciously set finite beings free for independent action is most
clearly seen in the case of human beings, who are defined by their
awareness of their own freedom and the need to act and to decide.
'Dialectic' does not mean paradox and contradiction, however. It is our
task to work to develop a theory of divine and human action that can
conceptualise *both sides* in ways that we can understand, rather than
developing the one in a way that excludes the other. (If our theology
does not meet these goals but appeals to mystery alone, we might as well
forego the activity of theorising and remain with the poetic, the
metaphorical or the ineffable alone, without the theoretical 'pretence'
intrinsic to the discipline of theology.)

Divine Action

There are a number of ways of conceiving the dialectic of divine and
human action in light of the doctrine of creation that shed light on
both without leading into contradiction. Divine action is most easily

understood from the perspective of the whole: God is the single source of all that exists, and through him the plethora of existing things receive the unity that allows us to call them 'the world'. The Greek philosophers and the Greek church fathers were thus right to link theology to philosophy by means of the question of the *source* or *origin* (*archê*) of the world. The strongest intuition that the theist has about God is that God is the source of and cause for the world. Nothing in contemporary physics stands in opposition to this fundamental intuition; as we have seen, physics may have the final say over the causes of events in the world, but it has nothing to say about the cause of the physical world as a whole, since it presupposes the existence of that world.

More difficult, particularly in our age, is to think of God as guiding the thoughts of particular individuals within the world, inspiring, encouraging and teaching. And more difficult still is to think of God as *breaking* the laws and patterns by which he created the universe in the first place. Each of these 'levels' of divine action represents a different theological locus: the first presents God as Creator, the second his role as providential guide of living beings, and the third that special form of divine providence traditionally associated with the working of miracles. The entire list forms a continuum, running from the divine role at the foundation of the universe (universal constants, the 'source' of the Big Bang) up to claims for highly specific interventions at concrete moments in human history. The entire question will concern us in some detail below. For now, it must suffice to note that God's role in salvation could be located at *any* of these levels. That is, the role of Creator could be viewed as *already* salvific, such that 'salvation' and the history of the universe are identified; or salvation could imply a more specific way in which God guides history, say toward a particular pre-planned conclusion; or salvation could be linked to individual miraculous interventions in history, such as the raising of the body of Jesus after his death.

CONCLUSION

What has emerged from this initial study is a clear sense of the importance of the doctrine of creation *for theological purposes*, above and beyond apologetic purposes such as defending belief in God 'against' scientific developments. The agenda for a broader discussion of the God/world relationship should thus not be set by the 'attacks' of scientific naturalists alone, but should flow naturally out of the belief in God the Creator. Thus Moltmann exhorts his readers not to go on distinguishing between God and the world, so as then to surrender the world, as godless, to its scientific 'disenchantment' and its technical exploitation by human beings, but instead to discover God *in* all the beings he has created and to find his life-giving Spirit *in* the community of creation that they share. This view – which has been called panentheistic (in contrast to pan-theistic) – requires us to bring reverence for the life of every living thing into the adoration of God.[59]

We found a highly particular story: the story of Yahweh's calling of a particular people in a particular time and place; the establishment of cultic practices and moral requirements; the construction of a narrative of sin, fall and redemption. Yet Yahweh's fierce jealousy – 'thou shalt have no other gods before me' – made this a system of cultic belief and practice that had to break the bonds of its particular context of origin. As it became clear that Yahweh's power (and moral demand!) was so great that other gods were 'as nothing', a more universal claim was made on his behalf: to be Creator of the heavens and the earth. Fundamentally, I have suggested, the doctrine of creation has to do with *the universality of the Israelite claims on behalf of its God.*

The same logic, we will find, applies to Christianity. The agency of God cannot begin with the fall and end with the salvific work of Jesus Christ, first in his incarnation and later through the life of 'the Christian cult', the church. If one took this approach, one would limit the significance of the story to Christians alone as those who explicitly acknowledge the state of sinfulness and the need for redemption. One would also subjectivise the entire process to a series of events taking place within the heart and soul of individual believers, or at most within a particular people. Indeed, exactly this has occurred in many modern approaches to theology, say since Schleiermacher. Modern theology has repeatedly shown the tendency toward a subject-based theology of salvation, bringing with it the concomitant dangers of subjectivising and privatising faith on the one hand, and of limiting its relevance outside of the Christian community on the other (perhaps with fingers pointed outward to the sinfulness of those on the outside). This is a point that Pannenberg has repeatedly stressed:

> If the dogmas of Christians are true, they are no longer the opin-ions of a human school. They are divine revelation. Nevertheless, they are still formulated and proclaimed by humans, by the church and its ministers. Hence the question can and must be raised

whether they are more than human opinions, whether they are not merely human inventions and traditions but an expression of divine revelation. Thus there arises once again, this time with respect to the concept of dogma, the truth question that is linked more generally to the concept of theology. . . .

Religious perceptions are thus exposed to the question whether they properly fulfil their function of bringing to light the infinite in the finite. In other words, the gods of the religions must show in our experience of the world that they are the powers which they claim to be. They must confirm themselves by the implications of meaning in this experience so that its content can be understood as an expression of the power of God and not his weakness.[60]

A renewed emphasis on the doctrine of creation and on the theology of nature has the potential to counterbalance such tendencies. Thus our opening treatment of the Hebrew Bible texts: the God who saves through Jesus Christ was first of all the God who created all things out of nothing. God is therefore either a God of universal relevance or is not God at all. The story of Israel, from the promise to Abraham through the Davidic period to the exile and return, would exist only in a corner of history, representing merely the story of a particular ancient tribe and its cultic god, if it were not for its claims on behalf of a creation out of nothing by the one God beside whom there is no other. If *this* God is concerned with redemption and salvation, the scope of his interest can be nothing less than humanity – no, nothing less than the universe as a whole. Exactly the same logic pertains to the story of Jesus, to his earthly ministry, and to the age of the Church that follows. This God who Christians assert acted in one particular individual (and in one particular institution) is the One who is responsible for all that is. It follows from this fact alone (in case there is any doubt) that the salvation question, and the teachings associated with it, are of universal import; the One who saves is also the One who holds all things in existence at every moment.

Equally important is what the theology of creation has to say about Christian truth claims. The 'postmodern' shift in recent years has placed great stress on abandoning 'meta-narratives' (Lyotard), on an 'internal realism' that limits truth claims to statements about the behaviour, beliefs and values of a given community. In this climate, it is very tempting to take the story of salvation and the associated teachings as being reports on the values of this given community that we call the church. But a particular starting point need not mean a limited import. To limit the potential significance of the biblical story runs afoul of *creatio ex nihilo*. The God who created the entire physical universe – all its matter and energy and all its lawlike regularities – out of nothing save himself must be at least as universal as each of the sciences that today express our knowledge of the physical universe. The doctrine of creation forces one, willingly or unwillingly, to speak of the universal significance, and hence the universal claim to truth, of the Christian message. God either is or is not the actual source of the entire physical universe. To assert that God *is* this source is to hold that the story of salvation is not merely a matter of internal transformation – a story told within, and hence definitive for the existence of, a particular community. The *justification* of claims to such broad significance may have become problematic in a postmodern context: theology finds a gap between the scope of its claims and the scope of the arguments it can make on its own behalf. This gap does and should inspire humility and a reassessment of the manner in which one proclaims what one believes. But we have learned from the doctrine of creation that it cannot alter the universal nature of *what* the biblical story is about.

NOTES

1. Paul Zimmerman, 'Can we accept theistic evolution?', in Henry M. Morris et al. (eds), *A Symposium on Creation* (Grand Rapids, MI: Baker Book House, 1968), p. 78.
2. W. Pannenberg, 'Kontingenz und Naturgesetz', in W. Pannenberg and A. M. Klaus Müller, *Erwägungen zu einer Theologie der Natur* (Gütersloh: Gütersloher Verlagshaus Gerd Mohn, 1970), p. 35.
3. Karl Barth, *Church Dogmatics*, III/1, trans. G. W. Bromiley (Edinburgh: T. & T. Clark, 1962), para. 41.2.
4. Bernhard W. Anderson, *Creation versus Chaos* (Philadelphia: Fortress Press, 1967), p. 49.
5. On the early dating of this psalm see Mitchell Dahood, 'Psalms II, 51–100', *Anchor Bible Commentary* (New York: Doubleday, 1958), q.v. Psalm 82.
6. Robert Butterworth, SJ, *The Theology of Creation* (Notre Dame: Fides Publishers, 1969), p. 36.
7. Claus Westermann, *Creation* (Philadelphia: Fortress Press, 1974), p. 4.
8. Wolfhart Pannenberg, *Systematic Theology*, Vol. 2, trans. Geoffrey Bromiley (Grand Rapids, MI: Eerdmans, 1991), p. 46.
9. See, for example, Robert John Russell, 'Philosophy, theology and cosmology: A fresh look at their interactions', Conference on Cosmology, Philosophy and Theology, Varenna, Italy, October 1996 (publication forthcoming): 'I would argue that science presupposes the contingency and

the rationality of the universe' (manuscript, p. 7). As supporting evidence Russell cites Ian Barbour, *Religion in an Age of Science*, The Gifford Lectures, Vol. 1 (San Francisco: Harper & Row, 1990), pp. 141–6, and his own 'Cosmology, creation and contingency', in Ted Peters (ed.), *Cosmos as Creation* (Nashville, TN: Abingdon, 1989), pp. 177–209.

10. Thomas Torrance, *Divine and Contingent Order* (Oxford: Oxford University Press, 1981), p. 26.

11. Claus Westermann, *Creation*, trans. John J. Scullion (Philadelphia: Fortress Press, 1974), p. 5, emphasis added.

12. See Vincent Brummer (ed.), *Interpreting the Universe as Creation* (Kampen, The Netherlands: The Kok Pharos Publishing House, 1991), p. 111.

13. See von Rad's essay in Bernhard Anderson (ed.), *Creation in the Old Testament* (Philadelphia: Fortress Press, 1984), p. 57.

14. Yehezkel Kaufmann, *The Religion of Israel: From its Beginnings to the Babylonian Exile*, trans. M. Greenberg (Chicago: University of Chicago Press, 1960), p. 60.

15. Leo Scheffczyk, *Creation and Providence*, trans. Richard Strachan (London: Burns & Oates, 1970), p. 6.

16. Thomas Torrance, *Divine and Contingent Order* (Oxford: Oxford University Press, 1981), p. vii.

17. Jürgen Moltmann, *God in Creation: A New Theology of Creation and the Spirit of God*, trans. Margaret Kohl (Minneapolis: Fortress, 1993), p. 74.

18. Leo Scheffczyk, *Creation and Providence*, p. 7.

19. As an example of this later 'reading back' even of the name Yahweh onto the earlier, pre-Yahwistic tradition, see, for example, the text from the Priestly tradition: 'And God said to Moses, "I am YHWH. I appeared to Abraham, to Isaac, and to Jacob, as El Shaddai, but by my name YHWH I did not make myself known to them"' (Exod. 6: 2–3). For a fuller discussion of the Tetragrammaton before Moses and this phenomenon of retrodiction, see G. H. Parke-Taylor, *Yahweh: The Divine Name in the Bible* (Waterloo, Ontario: Wilfrid Laurier University Press, 1975), especially Chapter 2.

20. Langdon Gilkey, *Maker of Heaven and Earth* (New York: Doubleday, 1959), p. 18.

21. Torrance, *Divine and Contingent Order*, p. 3.

22. See the interesting essays in Brummer (ed.), *Interpreting the Universe as Creation*, especially p. 107.

23. Leo Scheffczyk, *Creation and Providence*, p. 11.

24. Gabriel Daly, *Creation and Redemption* (New York: Gill & Macmillan, 1988), p. 22, italics added.

25. I leave aside for the moment the debate, initiated by Stephen Hawking, of whether quantum creation still allows us to speak of an initial moment of creation at all; we return to this topic below.

26. One should be more cautious about asserting that God *causes* the universe, since this suggests an external relation, the sort of effect one billiard ball has on another. Challenging the externality of God's influence on the world constitutes a major theme in the following chapters. See Michael

Welker, *Schöpfung und Wirklichkeit* (Neukirchen-Vluyn: Neukirchener Verlag, 1995), and Razinger's article in Josef Hofer and Karl Rahner (eds), *Lexikon für Theologie und Kirche*, 2nd edn (Freiburg: Herder, 1957–67).

27. Jürgen Moltmann, *God in Creation*, p. 206.

28. Gerhard von Rad, 'The theological problem of the Old Testament doctrine of creation' (originally published in 1936), in von Rad, *The Problem of the Hexateuch and Other Essays*, trans. E. W. Trueman Dicken (New York: McGraw-Hill, 1966), p. 142.

29. Robert Butterworth, SJ, *The Theology of Creation*, p. 26. Cf. von Rad in Bernhard Anderson (ed.), *Creation in the Old Testament*, p. 54, and Leo Scheffczyk, *Creation and Providence*, p. 4.

30. Cf. Bernhard Anderson, *Creation versus Chaos*, pp. 37ff.

31. Ronald A. Simkins, *Creator and Creation: Nature in the Worldview of Ancient Israel* (Peabody, MA: Hendrickson, 1994), pp. 90f.

32. Geoffrey Wainwright, *Doxology: The Praise of God in Worship, Doctrine, and Life: A Systematic Theology* (New York: Oxford University Press, 1980).

33. See Ernst Haag (ed.), *Gott, der Einzige: Zur Entstehung des Monotheismus in Israel* (Freiburg: Herder, 1985), especially the chapter by Erich Zenger, 'Das jahwistische Werk – ein Wegbereiter des jahwistischen Monotheismus?' (pp. 26–53).

34. Westermann, *Creation*, p. 31.

35. See von Rad's essay in Bernhard Anderson (ed.), *Creation in the Old Testament*, p. 63.

36. Langdon Gilkey, *Maker of Heaven and Earth* (New York: Doubleday, 1959), p. 20.

37. See Moltmann, *God in Creation*, p. xiv.

38. Brummer (ed.), *Interpreting the Universe as Creation*, p. 112, emphasis added.

39. Brummer (ed.), *Interpreting the Universe as Creation*, p. 107.

40. Dietrich Bonhoeffer, *Creation and Fall: A Theological Interpretation of Genesis 1–3* (New York: Macmillan, 1959), pp. 15f.

41. Bonhoeffer, *Creation and Fall*, p. 26.

42. Brummer (ed.), *Interpreting the Universe as Creation*, p. 125.

43. Harvey Cox, *The Secular City* (New York: Macmillan, 1965), p. 23. Westermann writes, 'The meaning [of creation] is that mankind is created so that something can happen between God and man. Mankind is created to stand before God' (*Creation*, p. 56).

44. See the exegetical and tradition-historical work in Ronald A. Simkins, *Creator and Creation*.

45. Westermann, *Creation*, p. 22.

46. Westermann emphasises that there is in fact 'a long series' of biblical creation accounts, extending through the whole history of the tradition; see *Creation*, p. 6. Thus it is illicit to speak of a single specification of the Creator/Creation relationship in the Hebrew Bible.

47. See, for example, the discussion of Genesis in the new series on Genesis, published by Bill Moyers as *Genesis: A Living Conversation* (New York: Doubleday, 1996).

48. Don Cupitt, *Creation Out of Nothing* (Philadelphia: Trinity Press International, 1990), pp. ix–x. Cupitt's approach is echoed in the highly influential recent book by Karen Armstrong, *The History of God: From Abraham to the Present, the 4000-year Quest for God* (London: Heinemann, 1993).

49. See Labuschagne's essay in Vincent Brummer (ed.), *Interpreting the Universe as Creation*, quote p. 124.

50. Bonhoeffer, *Creation and Fall*, p. 41.

51. See Pannenberg, *Anthropology in Theological Perspective*, trans. Matthew J. O'Connell (Philadelphia: Westminster Press, 1985).

52. See Michael Theunissen, *The Other: Studies in the Social Ontology of Husserl, Heidegger, Sartre, and Buber*, trans. Christopher Macann (Cambridge, MA: MIT Press, 1984).

53. See Philip Clayton and Steven Knapp, 'Ethics and rationality', *American Philosophical Quarterly* 30 (1993): 151–61.

54. See Plessner's concept of 'openness to the world' (*Weltoffenheit*), for example in Helmut Plessner, *Die Frage nach der Conditio humana: Aufsätze zur philosophischen Anthropologie* (Frankfurt: Suhrkamp, 1976). See also Plessner's *Philosophische Anthropologie: Lachen und Weinen, Das Lächeln, Anthropologie der Sinne*, ed. Gunter Dux (Frankfurt: M. S. Fischer, 1970) and his *Conditio humana* (Frankfurt: Suhrkamp, 1983).

55. Bonhoeffer, *Creation and Fall*, p. 97.

56. '. . . while some may evaporate in their own superstitions and others deliberately and wickedly desert God, yet all degenerate from the true knowledge of him. And so it happens that no real piety remains in the world . . . One may easily grasp anew how much this confused knowledge of God differs from the piety from which religion takes its source, which is instilled in the breasts of believers only' (Jean Calvin, *Institutes of the Christian Religion*, 2 vols, ed. John T. McNeil (Philadelphia: Westminster Press, 1960), Vol. 1, pp. 47, 50).

57. See Richard Dawkins, *The Selfish Gene*, 2nd edn. (New York: Oxford University Press, 1989), and – in defence of the long-term survival value of altruistic behaviour – Melvin Konner, *The Tangled Wing: Biological Constraints on the Human Spirit* (New York: Holt, Rinehart & Winston, 1982).

58. See Pannenberg, *Anthropology in Theological Perspective*.

59. Moltmann, *God in Creation*, p. xi.

60. See Wolfhart Pannenberg, *Systematic Theology*, Vol. 1, trans. Geoffrey W. Bromiley (Grand Rapids, MI: Eerdmans, 1988), pp. 9f. and p. 167, as well as the entire third chapter, 'The reality of God and the gods in the experience of the religions'.

3

CHRISTOLOGY AND CREATION: STRUGGLING WITH THE PARTICULARITY OF THE CHRISTIAN STORY

We have examined the notion of God that arises out of and is presupposed in the doctrines of creation and nature. In the previous chapter we concentrated on the Hebrew documents and the Israelite theological tradition, under the assumption that that there is a circular, dynamic relationship between the scriptures of a tradition and theology's subsequent 'systematic' reflection on them.

Now that these doctrines have been presented, at least in overview, it is time to undertake a new task: to bring them into explicit connection with the narratives concerning the life and death of Jesus Christ and with theological reflection on their significance. The goal is not to produce a complete christology, a systematic doctrine of Christ, since this will be done in another volume in the present series. Rather, our task is twofold: first, we must observe how the Hebrew understanding of God and God's relation to the world is transformed when it is brought into contact with the claims specific to *Christian* theism. Second, we must wrestle with the 'stumbling block' of particularity inherent in Christianity. Both Jewish and Christian practice involve highly specific beliefs about God and God's will for humanity. But the Christian need to incorporate Christ into the being of God involves a break with Israelite monotheism and adds a further set of highly specific claims that compound the problem.

But why 'stumbling block'? Did this study not commence with the recognition that a postmodern or 'post-foundationalist' approach to knowledge does not need to begin with the universal and deduce the particular? Instead of universal foundations, we began with the particular account of God in the Israelite tradition (though we did find there beliefs, narratives and symbols of universal import). The postmodern shift, the opening chapter argued, no longer forces one to eschew the concreteness of individual narratives and the particularity of the Jewish and Christian beliefs. Much more, it embraces the fact that religious practices and beliefs are located and embedded within particular cultural and historical traditions. One does not first need a prior justification in order to introduce a religion's salvific narrative and subsequent reflection on the significance of that narrative.

As it turns out, however, the very theory of knowledge that frees Christian theologians to speak of the kerygma of Jesus Christ without embarrassment also changes the discussion with science in subtle ways. At the level of concrete narrative there is no conflict because there is (little or) no contact. It is only as theological reflection moves to a more systematic level – as it speaks about the direction of history, or the process by which the world originated, or the best explanations for particular phenomena in our environment – that points of contact begin to emerge and the task of harmonisation demands attention. As the sciences have come to form a more and more significant background context in the doing of systematic theology, the pressures of this emphasis on the 'universal' have increased. Frequently, the systematic theologian now finds herself torn between the specificity of the biblical texts on the one hand and the general scientific (or philosophical) debates on the other. (Interestingly, the dominance of metaphysics in the pre-modern era placed very similar pressures on theology.) In a word: the discussion with science has tended to problematise the highly individual and concrete focus of Christianity at the same time that the postmodern context should have removed its onus.

One finds three types of response to this conundrum. Some have maintained that what really matters about Christian theology (or *all* that really matters) is not the particular story but its meaning at this universal level. The concrete narratives then become subordinated to their universal import, and Jesus becomes an instance of a broader type or category. Others have concluded that story mattered (past tense), but that philosophy (or science, or comparative religions) have now lifted us beyond such provincialism. The story cannot be kept even as an example of universal truths; at best it is a myth, at worse a diversion from the (general) truths we have come to hold. Finally, some Christian theologians have reacted to these conflicts by stigmatising all intercourse with science (or philosophy or other religions). If such dialogues question or take attention away from the life and death of the God/man Jesus Christ, then Christians should avoid them.

My own response is to suggest a division of labour. There is need for biblical scholars who elucidate the teachings and actions of Jesus Christ and of the various biblical writers' responses to them. And there is need for detailed work on the theological loci unique to Christian thought: on christology and ecclesiology, on the doctrines of redemption, salvation, sanctification and the sacraments. At the same time, systematic and philosophical theologians – those who make use of conceptual and metaphysical frameworks in explicating the broader implications of Christian thought – cannot help but engage with philosophy and the sciences. Now the greater stress in the present work will obviously fall on the second category. This might give rise to the impression that the particularity of Christian theology has been left behind (as in the first two responses above), replaced by a generic theology of 'theism-in-general'. But this would be a misinterpretation. The debate in later chapters may indeed be carried out primarily between, say, personalistic theism and scientific materialism. But the particular theism that motivates the discussion for the Christian theologian continues to be the Christian doctrine of God. The purpose of the following reflection is to underscore the unique understanding of *who* this God is who, according to Christian proclamation, is active in the world.

INDISPENSABILITY OF THE KERYGMA

Christianity, as its name implies, begins with the story of Jesus Christ. Its testimony is that in this person God acted *pro nobis* – for individual humans, for the community of those who bear that name and, universally, for humanity and for the universe as a whole. How is one to determine the viability of this story and assess its meaning? The personal element in *pro nobis* suggests that one consult one's own experience and the experience of the church. But, as Hans Frei rightly saw, appealing to one's own individual consciousness, or even the corporate consciousness of God – the moves that were definitive for Friedrich Schleiermacher, the 'father of modern theology' – is not sufficient to ensure the distinctiveness of Christian belief.[1] That distinctiveness must finally flow from the story itself and from the claims made on its behalf by believers.

Should the distinctiveness of Christianity be preserved? Many would say that it is better in today's context of multiple religious traditions to emphasise universal commonalities among the world's traditions. For example, perhaps we should concentrate on the parallels between the 'Christ consciousness' (Schleiermacher) and the 'Buddha consciousness', seeing the two as expressing similar spiritual insights, albeit in different cultural trappings. Without careful study of the various religious traditions involved, and without the development of an appropriate inter-religious hermeneutic (or categorial system), it is impossible to know to what degree they all involve the same, similar, complementary or incompatible religious beliefs. Equally difficult to decide is whether it would be desirable or undesirable if all religions spoke with the same voice (my own view is that it would be both desirable and undesirable, but for different reasons).

It does seem clear, however, that, unlike metaphysical systems, most religions are highly concrete. Each one relies in an essential manner on particular stories, symbols, persons and historical events. There are therefore *religious* reasons for seeking to preserve the particularity and uniqueness of particular religious traditions rather than subsuming them under overarching, trans-religious categories or metaphysical frameworks. If this is correct, then it poses for us a distinctively theological task: how to conceive – to understand, to ground and perhaps to guard – the particularity of Christian (or Jewish or Hindu) belief and practice in contradistinction to that of other traditions.

The Christian theologian can therefore no longer present Christianity as merely one particular instantiation of more universal religious truths. The focus will instead be on what are the distinctive claims that set it apart from other traditions, and in particular from the other main monotheistic traditions, Judaism and Islam. Clearly, the beliefs that are distinctive to Christianity are crucially related to the person and work of Jesus Christ. It was again Hans Frei who insisted that Jesus is nothing apart from the story of his life.[2] That narrative, in turn, is not available apart from the written form in which it has been transmitted in the Hebrew and Christian scriptures. Hence these scriptures, and the story they tell, play the crucial role in determining Christian particularity.

Of course, a Christian theology that appeals unapologetically to the New Testament texts need not contradict appeals to either private or corporate experience. For the writings about Jesus have informed individual and ecclesial Christian experience from almost the very beginning, and these experiences have in turn repeatedly (and inevitably[3]) informed interpretation of the texts. A similar hermeneutical circle obtains on the level of doctrine: the history of Christian theology has been guided by those writings (indeed, theologians helped to constitute them in the first place through the early councils), and theology in turn serves as the screen – hopefully one that is not too opaque! – through which most readers approach those documents today.

Before we sketch the implications of the scriptural narratives regarding Jesus the Christ, it is good to emphasise again the starting point in the Christian kerygma: Jesus Christ lived and died on behalf of the whole world. Through his life the Father reinstated a right and whole relationship of humanity with himself. Christ's atoning death might have been a futile (though noble) gesture if it were not for the fact that, according to Christian proclamation, the one Jesus called 'Father' attested powerfully to the unique status and actions of Jesus. In an action that symbolised the Father's acceptance of Jesus' self-sacrifice and testified to the significance of his death, the Father raised Jesus from the dead. Through this act, Paul writes, God caused Jesus' propitiatory action to be efficacious, using it to establish a way for reuniting humans to himself: 'For if by the transgression of the one the many died, much more did the grace of God and the gift by the grace of the one Man, Jesus Christ, abound to the many' (Rom. 5:15). The one who accepts this action of Jesus on her behalf is *en Christo*; she communes with God in and through Christ, and lives for God by the power of God's Spirit.

IMPLICATIONS OF THE CHRIST EVENT

If this were a book on christology much, much more would need to be said about the person, work and teachings of Jesus. In this brief chapter we must unfortunately limit ourselves to the barest outline of christology. How is the Israelite understanding of God transformed within Christian theology by accepting the story of Jesus and its most basic entailments? At least seven central doctrines must be mentioned here, since they undergird the systematic reflection in the following chapters. It may be that much of the debate with the sciences involves general questions involving the existence of God and the nature of divine action. Yet where *Christian* theism is intended, these are the sorts of specific features that lie in the background.

1. In the first place, the story of Jesus represents the primary recognition for Christians of where God has been active in history and of what the concrete actions were with which he was involved. Basic to the Christian narrative of divine action is the belief that God has a concern for human beings, their beliefs, and their acts – both in relation to him and in relation to each other (the doctrine of divine love or benevolence).

2. Second, the Christian narrative expresses the belief that God is also concerned with the distance of human beings from himself, with their tendency to disregard (or flaunt) their nature as created beings and to seek to make themselves not finite but infinite, becoming their own source of meaning and being (the doctrine of sin).

3. God would not be God were he merely to have *responded* to Jesus' redemptive initiative on humanity's behalf, as if the whole event had emerged as the happy coincidence of a human intention with the divine intention, apart from any agency of God the Creator. Instead, the Father must actually have initiated and must have been responsible for the actions that Jesus would accomplish on humanity's behalf (the doctrine of providence, the sending of the Son by the Father).

4. Put differently, the work of Christ the Son must be understood to have been the work of God the Father as well. Even if the particular functions or roles that the two played were different, the work was so fundamentally the work of both that they must be understood as two equi-divine persons, God in two Persons. Likewise, the work of the Spirit who mediates between them, who glorifies both Father and Son and who completes the Son's work as his continuing presence on earth, suggests that the Spirit is a third irreducible unit in the salvation story, an independent Person within the one Godhead (the doctrine of the Trinity).

5. This means that Jesus Christ must be understood as the full union of the second Person of the Trinity with the human individual Jesus (the doctrine of the Incarnation).

6. God the Father accepted and honoured the salvific actions of Jesus (the doctrine of salvation or redemption). Likewise, he continues to be present with those 'who once were far off' (Eph. 2: 13) and who were redeemed by the work of Christ (the doctrine of the church). The Father will send the Spirit in the name of Christ (John 14: 26) so that he can continue to be present with his people (the doctrine of the Holy Spirit). The Spirit continues the work begun by Christ (the doctrine of sanctification).

7. God intends to complete the work begun in the life and death of Jesus and now carried on by the Spirit. Indeed, for God to be God he must finally be able to perfect or bring to completion the work begun in the life and death of Jesus. Bringing about this broader purpose, 'the oneness of all in Christ', at the end of history is the consummation toward which God and all creation strives (the doctrine of eschatology).

Much more can be derived from the basic story of Jesus, enough to fill shelf upon shelf with material; the foregoing represents only the briefest allowable sketch of the core theological beliefs that the tradition has taken to be entailed by the Christian story. For some readers, it will constitute the absolute minimum expression of Christian belief (indeed, some will probably hold it to fall short of that minimum until it is supplemented by, for example, a more robust doctrine of scripture, of the church and sacraments, of Christian obedience). For others, some or all of these doctrines will simply be 'non-starters', obviously untenable in today's context. For them, additionally, the focus in the present chapter on this particular narrative will be uncomfortable to the point of unacceptable. Much better, these theologians will reply, to seek formulations that include all religious traditions simultaneously and without distinction, and to let as many as possible of the traditional doctrines go. I believe, however, that this move is mistaken. The doctrines just summarised, which represent a sort of 'minimal Christian theology', are the beliefs that have constituted the Christian tradition. Here it is that we must begin – wherever we then go from here.

The pill is perhaps not quite so bitter to swallow as it may seem to some. At least three comments are possible about what it means to advocate a minimal Christian theology of this sort.

1. The *form* that these doctrines take, what they mean and what significance they have, is not written in stone. The formulations have developed and metamorphised through time; they have needed revision before and will need it again. They represent, as the German theologians say, *Denkaufgaben*: tasks for thought.[4]

2. To tell one story is not, in the post-foundationalist context, to pronounce all other stories false. Nor does making a truth claim mean that one's present formulation is the final formulation, such that all others are excluded. In this work I neither defend or attack the project of a meta-tradition (e.g. a world theology) that would seek to do justice to the concerns of multiple religious traditions. Christian theology – understood in this chapter as the attempt to spell out the logic of the Christian story – may or may not be willing to see itself as one particular expression of spiritual insights shared with other traditions.

3. Finally, a tradition is a living thing, not a static object to be defined once and for all. Where the reader finds particular doctrines or formulations troubling, she is encouraged to wrestle with them, to find better formulations, to argue for the rejection of those she

finds unredeemable. Stating the Christian story using its traditional parameters is not a weapon to be wielded against misguided individuals in order to exclude them. It is a shared challenge, for those who choose to bear the name Christian or who find themselves associated with its beliefs and practices, to find formulations that do justice both to the Christian tradition and to the intellectual context in which it finds itself today.

THE TRINITY

The previous section sought to show how much is revealed for the Christian theologian about the nature of God by God's involvement with the events of the life, death and resurrection of Jesus. The next step is to reflect further on who this Jesus must have been to have played the role in salvation history attributed to him by the biblical texts. It does not seem sufficient, I argued, that he would have been an individual human being, unprompted by God, who just managed to 'key into,' comprehend and communicate to humanity the fundamental plan and nature of God like a computer hacker breaking a Defense Department code. If Jesus *were* some kind of 'spiritual genius' who managed to achieve an accurate knowledge of God's plan on his own without tutoring, and whom God rewarded with resurrection like an 'A' grade for a course, then God's divinity, his contribution to humanity's knowledge of him, would be cast into question. Now God might have moulded Jesus in some way, taking a spiritually astute individual and guiding him into spiritual wisdom like a sports trainer might work with a gifted athlete. But this kind of picture also does not seem sufficient for God's revelatory action in the Jesus-event, since it leaves the initial emergence of Jesus to chance. Also, it is difficult to understand how a 'mere human' would have the status, and thus the authority, to re-establish a right relationship between God and humanity. Would it not rather be that whatever agent is involved in such an act would need to participate in some way in the divine nature itself?

In this line of argument lies the motivation for the theological proclamation of Jesus as divine. Jesus must somehow have shared in the divine nature itself in order to have played the crucial role in salvation history that is ascribed to him by traditional Christian belief. Now the last chapter has already discussed what it means to say that God is personal. How should this personal nature be conceived in order to incorporate the story of Jesus reviewed above? For Jesus clearly acted as a person and not as an automaton in this narrative; his actions involved his own will and intentions in a specific and irreducible manner. If our context were pantheism, then it would be easy to spread the divine spirit around, assigning little bits of it (like fairy dust) to all parties involved, to some a little more and to some a little less. Such a treatment of Spirit will not work for theism, however, since theism in all its guises understands there to be a real distinction between the Creator God and the contingent beings of his creation.

If we are to do justice to the Christ event, it seems that we will have to speak of at least two centres of personal activity within the divine being, yet without running the two centres of personal activity together and without equivocating on the word 'divine'. According to the logic of the Christian story, Jesus Christ was a divine person; hence the God to whom he spoke must have been a divine person in some ways distinct from him. For example, Jesus spoke of this God as 'Father' and spoke of himself as 'Son' in relation to the Father. Thus we must think of the Godhead as including both Father and Son. The divine must include at least two divine persons, with 'God' as the summary word expressing the nature shared by these two persons. With this move we have encountered the first supporting arguments for the Christian doctrine of the Trinity.

The importance of this line of reflection for Christian thought cannot be overestimated. Left behind is the one God in awesome splendour, whether understood as unlimited power (the sheer force that lies at the origin of the universe; the *mysterium tremendum et facinans*[5]) or understood as pure thought thinking itself (*nous noetikos*) in the fashion of Aristotle's doctrine which was so influential on Scholastic theology. Instead, within the Christian tradition God is defined from the outset as *community*, a community of persons engaged in eternal fellowship. Of course, anything that we say about this community of divine persons is inevitably an anthropomorphism – whether one projects one's human understanding of persons directly onto God, or whether one moves from God's self-revelatory actions in human history (the economic Trinity) back to God's central nature (the immanent Trinity). Nevertheless, even when extreme care is taken to avoid anthropomorphism, much can be derived from the community notion, as Joseph Bracken has compellingly shown in a recent book.[6] It is God's nature to be social, to engage in interaction, giving, receiving. Consequently – and this inference is extremely significant – it is God's nature to be personal. Philosophers have long found it difficult to conceptualise the personal nature of God within the context of metaphysics, that is apart from the trinitarian context. Not infrequently, the difficulties led thinkers (most notably Hegel) to postulate the necessary existence of the world, for only in light of the world as God's 'other' does it appear possible for God to be personal and to enter into real relations with an other in the sense that seems to be required for an application of the predicate 'personal'.

Most importantly, the trinitarian notion allows for God's nature to be understood as love. St Thomas had no difficulty in introducing Goodness as one of the eternal features or 'transcendentals' of the divine nature. After all, benevolence seemed to be a divine perfection that could be inferred from the list of perfections that must pertain to God as *infinita perfectio* (infinite perfection). Nonetheless, when reading the philosophical discussions of the nature of God one often has the impression that something is missing when authors 'infer' the benevolence of God. The resulting qualities just do not seem sufficiently similar to the connotations of the term 'love' that we are familiar with in human contexts and in the New Testament. Presumably this is because the predicate *love* requires a genuine mutuality and reciprocity, a giving and receiving among multiple persons. Philosophical monotheism in the strict sense – the God of Infinite Perfection – does not lend itself naturally to such contexts of mutuality (except perhaps by including a necessary world as God's other, but there are well-known difficulties with such a position).

By contrast, the core of the trinitarian notion of God – and here I follow Wolfhart Pannenberg[7] – is that the one God consists of three separate persons. The Son eternally gives himself to the Father, not claiming Godhood but seeking to glorify the Father. The Father accepts this movement, this love, on the part of the Son who proceeds from him and in return continually offers full Godhood to the Son. The Spirit mediates between the two, glorifying the Son and constituting the presence of the Father. One does not have to be convinced by all the details and complicated formulations of the trinitarian relations – and especially not by the quasi-necessity claims made in some of the accounts[8] – to sense the strength of this model of the divine life as one of mutuality, giving and receiving. God-as-community is certainly a far cry from the isolated God of Aristotle, conscious of nothing but itself, never even knowing that it gave rise to an entire universe which exists as a continual imitation of and quest for the Unmoved Mover who is its *telos*. It is also no coincidence that Karl Barth has stressed the features of the love of God and the freedom of God as central to the being and reality of God.[9]

Thus it is that the God of Christian theology is far removed from generic or philosophical theism. No process of inference can lead one from the God of Infinite Perfection to the God of Abraham, Isaac and Jacob, nor to the God and Father of Jesus Christ. Philosophical theism, as we will see in the next chapter, does provide conceptual criteria for theological doctrines of God and (sometimes) helpful models for the systematic theologian. Ideally, Christian trinitarian theology can incorporate the best insights of the philosophical doctrine of God. But no reduction of the one to the other should be expected.

What holds for the nature of God in general pertains also in the case of each of the specific attributes of God. In philosophical discussions, for example, omnipotence is defined as absolute power, the power to do anything, a power limited only by the bounds of what is logically possible. But 'omnicausality' cannot possibly serve by itself as the definition of omnipotence within Christian theology. For the God of Christian theism, as Barth has stressed, omnipotence must be conceived as the omnipotence of love. As such, it cannot possibly destroy or even conflict with the freedom of the creature, in so far as God expresses his power toward creatures precisely through his free and loving creation and election of them.[10] The power of God's will (omnivolence) is not an abstract calculus of power, as in Descartes' discussion of God's ability to create the 'eternal truths', but rather pertains to God's will as it has actually been revealed in his actions toward humanity. Omnipotence, Barth suggests, is a 'moral power' (ibid., 526ff.). Likewise, omnipresence involves both God's creation of the space in which humans can exist and meet him, and his loving presence with all of creation at all times. Similarly, again following Barth, omniscience is not primarily the ability but first of all the fact of God's fully knowing (and fully caring for) all parts of his creation at all times. For the Christian theist, then, omnipotence and the other attributes must be defined christologically: the God who has acted in these ways in Christ is the very same one to whom we ascribe the fundamental philosophical predicates and 'perfections'.

THEOLOGY, PHILOSOPHY AND HISTORY

There would *be* no Christian theology (in distinction from philosophy in general) if it were illicit to think of the nature of God in light of God's self-revelatory acts as reported in the Christian story, using the Hebrew and Christian scriptures and the history of interpretations incarnated in the life and thought of the church. Yet the dependence goes both ways. Following Tillich and many others, I have allowed broader categories and questions to inform work within theology proper. At the same time, it would be a mistake, having once supplemented generic theism with the richness of the Christian categories, henceforth to move solely on Christian terrain and in Christian terms. Instead, having once made the religious move to 'God with us', and having elucidated the notion of God by means of Christian symbols and the Christian narrative, we need then to draw on every possible source in order to gain the fullest possible knowledge of what these claims might mean. Here I am again indebted to Pannenberg: theological claims are viewed as hypotheses through which we aim to gather more information about the true nature of the world. Like hypotheses, they remain dependent on final verification from the universe itself and its history.[11]

One might think, by analogy, of the great ontic movement in the *Summa Theologiae* of Thomas Aquinas. In that work, perhaps unequalled in the entire theological opus, St Thomas moves from the basic features of our human experience (classically, in the five *viae* of I/1, q. 2 art. 3) up to the nature of God (most of I/1), down through the salvation-historical movement of incarnation, redemption and the establishment of the church, and finally up again in the movement of humanity toward God (sanctification) and of the entire creation to a final reconciliation and completion.

What St Thomas has described *ontically* must also occur *epistemically* for the Christian theologian and believer. We know that no natural theology will lead one to the trinitarian Christian God, even though the evidences and indications drawn from human experience remain crucial as a starting point (this too was Thomas's assumption at the opening of the *Summa*). In a post-foundationalist context the Christian narratives give one a vocabulary for speaking of the nature of God, of God's self-revealing acts in Christ and in history. Nonetheless, doing theology in terms of the Christian story is not the final step. Like Aquinas, the theologian must then complete the movement back outward again to encompass the wealth of the total data available to her. She must 'redeem' the total set of evidence about the world as she attempts to incorporate it within a fully articulated theological vision – hence the treatment of God's activity in the world *in light of* contemporary science.[12] The noetic question, just like the ontic, will not be consummated until the final coming of God. But the task of incorporating 'all knowledge', like the ontic task of 'unit[ing] all things in [Christ], things in heaven and things on earth' (Eph. 1: 10), is an indispensable part of the broader theological task.

It is tragic that Karl Barth, from whom theology has otherwise learned so much in this century, not only neglects this broader task but consciously eschews it. One cannot help but note his relief when in the opening pages of III/1 he 'realises' that he does not need to know anything whatsoever about science, since all that is required of the theologian is to confess God as Creator: 'There can be no scientific problems, objections or aids in relation to what Holy Scripture and the Christian Church understand by the divine work of creation.'[13] From the claim that we know creation as an item of faith – a claim that I do not dispute – Barth moves on to the conclusion that we know creation (only?) from scripture as God's witness to Jesus Christ:

> The statement concerning creation cannot be anything but an *articulus fidei* because (1) its assertion of the reality of the world . . . and (2) its grounding of this reality in God are possible only as a statement of faith, and because (3) it is determined in all its elements (subject, predicate and object) by the linguistic usage of Holy Scripture and the content of the terms employed . . . [Thus Barth's question is:] How is it that the Christian Church – whether or not it is illuminating and pleasing to itself and the world, and in spite of every objection and contrary opinion – can publicly confess this work of God the Father, the creation of heaven and earth, as truth and indeed as absolute and exclusive truth? . . . What is it that permits and commands us . . . to treat this truth as the presupposition which has to precede all other presuppositions, axioms and convictions, and the effectiveness of which can be counted on in every conceivable connexion? (22f.)

Barth's limitation, I suggest, is unnecessary and damaging to the theological task.[14] For the Christian theologian, knowledge of Jesus Christ may well be a necessary condition for knowing creation; by no means, however, can this knowledge constitute the *sufficient* condition for knowing creation. Contrast Barth's otherwise masterful treatment of creation with the treatment of creation in Pannenberg's second volume of the *Systematic Theology*. Pannenberg's explicit goal is to provide a trinitarian understanding of creation, and he spells out the role of Father, Son and Holy Spirit in creation, using the resources of both historical and systematic theology, in a way that leaves no doubt as to the absolute centrality of the theological context. Nonetheless, Pannenberg also draws brilliantly on central concepts from science and anthropology[15] to provide a much fuller account of what creation means than would otherwise have been possible. The theologian's understanding of spirit, of power, of space and time, and of contingence and infinity is enhanced by the scientific and philosophical explorations that are fundamental to Pannenberg's overall analysis. The construction of the present volume, with its theological exploration of problems and data from the sciences and philosophy, is meant to express a similar commitment, a commitment visible, for example, in the movement from biblical data to theological reflection to less traditionally theological themes in the final chapters.

Note that one can break with Barth here and nonetheless concede that he is right in an important way about the status of theological knowledge about creation. Creation is neither myth (in the common sense of fiction) nor a historical account in the way in which historical records are generally written (*Historie*). It is 'prehistory', as Barth writes: the account of creation refers back on the one hand to events that took place in space and time, to actions that really occurred. On the other hand it concerns an action that occurred before recorded history – and necessarily so, since there *could not have been* human observers to this particular divine action! Note that something similar is surely true of the resurrection: although it seems clearly inadequate to call the resurrection a helpful fiction that expressed the early disciples' trust in God or their belief in the infinite resources of humanity, it also seems unlikely that a scribe, sitting next to the body of Jesus from the moment of his death until the resurrection was completed, would have been able to record (or videotape) a resurrection. A resuscitation, perhaps, but not a resurrection – since resurrection is a theological concept and a theological event.[16]

This Barthian insight remains remarkably helpful today as Christians attempt to understand the status of Christian truth claims in a pluralistic context and with full realisation of the epistemic limits on their speaking. Whatever we write or say has the status of a saga, tale or story: 'In addition to the "historical" there has always been a legitimate "non-historical" and pre-historical view of history, and its "non-historical" and pre-historical depiction in the form of saga'.[17] The biblical accounts reflect the historical conditionedness and cultural influences that mould the authors' (and our) particular ways of viewing and interpreting nature and creation. Yet what they and we refer to is not – and cannot be viewed merely as – a fiction, however limited their (and our) perspectives. All believers make truth claims, and Christian truth claims have traditionally referred to actions and realities that stand above every cultural and historical perspective. Drawing on scientific and philosophical knowledge does not compromise this point. Today, when theorists of science are so clear on the limitations of scientific and philosophical knowledge, appealing to them is not a sign of the hegemony of philosophy but rather yet one more reminder of the limited nature of all human knowledge. It is therefore possible to carry out the theological project in dialogue with the sciences and philosophy in a way that was inconceivable to Barth, who still lived under the myth of scientific objectivity and philosophical absolutism. The significance of this shift for the very fabric of theology and for its fundamental methods is only beginning to be understood.

EMBRACING ONE PARTICULAR AMONG OTHERS

We have commenced our project in these opening chapters by considering the Hebrew and Christian doctrines of creation and the God/world relation. To start in this way expresses a commitment to the particularity of the starting point for Christian theology. One begins with a general question, the search for the locus of divine activity and revelation in the world. It is a question shared by numerous religious traditions, each of which turns to a different set of scriptures for its answer. In today's context, the awareness of this plurality of traditions, and the knowledge that humans are divided concerning the results, gives special meaning and urgency to the Christian affirmation. What characterises the Christian believer is her focus on the narrative of Jesus Christ – including its roots in the history of Israel and its subsequent incarnation in the church – as a focal point of divine activity. She may even hold that God has been uniquely present to humanity in the person and work of Christ. At the same time, she knows that other believers make similar claims for the history of Judaism, or for Buddhism, or for others. She also knows that still others argue that God is equally visible in *all* the world religions,[18] and yet others argue that God is visible in none of them.

I cannot imagine that this project – the quest for possible divine influence on the human race and its history – could be pursued today, especially within the academic context, apart from this horizon of (at least) the major religious traditions. At the same time, comparative efforts require insightful theologies from within each individual tradition. Scholars specialising in particular religious traditions thus do a double service: they both represent their own traditions for 'outsiders', helping others to recognise the essential features and claims of their tradition, *and* they assist 'insiders' in a particular tradition to see more clearly what is the 'logic' of their own tradition.[19] Thus, even in light of a comparative and pluralistic perspective, the detailed *and internal* consideration of the features of the Christian story remains indispensable. To put the point differently: pluralism does not entail a 'suspension of belief', a belief-free neutrality that sits calmly in judgement over all traditions of belief (this was the major error of the Enlightenment). Even the comparativist perspective must thus stress what may be learned of the divine nature through the distinctive characteristics of Christianity (as also Judaism, Islam, Vedantic Hinduism, etc.). In Christianity (as traditionally understood), the distinctive claim is that God himself was present in the person Jesus in an ontologically unique way. Hence the urgency of the quest to understand what belief in the Incarnation means and entails. This means that something very like the traditional task of systematic theology is still called for: a re-telling of universal history and salvation history using the terms and language of the Christian tradition.

Gone therefore is the old Enlightenment injunction that only 'external' or universally accessible (i.e. philosophical) language be used to speak of God. The line of argument has pointed us instead toward fully concrete and historically bound symbols and narratives of divine action. At the same time that systematic theology has been rescued from rationalism, however, its broader context and self-understanding has been transformed. That context is now the plurality of world religions, which means, at least, a plurality of beliefs about when and how God has become manifested in history (and, in some cases, a deeper questioning about the very notion of God). For those who reflect on religious phenomena at the close of the millennium, there is no escape from pluralism. Theologians seek to understand and elucidate the concepts of God and world basic to the Christian message; yet they do so within the arena of diverse and often conflicting truth claims from the various world religions. Theology cannot escape the truth claims of Christianity – following Jesus can never be merely a matter of taste – but nor can it act any longer as though they were the only truth claims in town.

WHO BEARS THE BURDEN OF PROOF?

In the previous paragraphs we have looked at the pluralistic or multi-religious context within which theology must be done today. There are reasons to be optimistic about the resources that traditional Christian theology brings to the contemporary world. But if these resources are to be utilised, they must be focused on the questions that are being asked today and they must be used in a manner consistent with that context. One issue that emerges again and again in the discussions – with science, with naturalism, with the world religions – is the question of who bears the burden of proof. Do Christian theologians first need to show that atheist alternatives are false? Conversely, do atheists need to show the untenability of Christian belief in order to be justified in their positions? If neither of these conclusions is correct, is everyone justified in holding whatever beliefs they find themselves holding, regardless of the arguments brought against their position or in support of alternative positions? In the present work I have by and large circumvented debates about *religious epistemology*, the question of the criteria and justification for religious belief.[20] At this point, however, we should pause to give it at least passing attention.

A Brief History of the status quaestionae

Recent Christian thought has been characterised by a rich variety of positions which provide sophisticated reasons for answering 'no' to the first question in the previous paragraph. One recalls Karl Barth's famous 'No!' to Brunner and to natural theology: because God's revelation breaks in from above like a lightning bolt, it stands in no continuity with the world or with the world's standards for knowledge. Hans Frei, the founder of the Yale School, was inspired by Barth when he turned to the category of *narrative* to address the sceptical challenges and the tendencies toward natural theology that he found in liberal theologians in the 1970s. Frei argued that the biblical texts bear their own meaning; the text is to be interpreted *out of* the text, and not upon the 'foundation' of any philosophical system or set of scientific beliefs.[21]

Building on Frei's earlier work, George Lindbeck then argued in *The Nature of Doctrine* that the church's context of worship and practice provides the only necessary horizon for understanding and testing Christian truth claims. The truth of any given statement is to be judged in a 'cultural–linguistic' fashion, using as backdrop the context of Christian practice and the scriptural story taken as a whole – and definitely not through any statement-by-statement confrontation with scientific or philosophical truth claims.[22] Similar arguments have been made with a more explicit reliance on twentieth-century philosophers. Thus Lindbeck appeals to Wittgenstein in arguing that truth claims must be understood relative to a particular 'form of life'. Christian language and practice represent one such form of life, and theology's task is merely to spell out the 'grammar' that underlies the way that Christians use Christian language in Christian practice.[23]

Theologians have been able to find additional philosophical allies in their efforts to remove Christian language from broader scrutiny. We have already mentioned Wittgenstein's theory of language games. Lindbeck also makes use of anthropologists such as Peter Winch and Clifford Geertz, who argue that each culture or subculture represents a form of life of its own and cannot be adequately understood or conceptualised

in any terms save its own. Theorists of narrative have encouraged theologians to treat the biblical texts as narratives that must be interpreted
in their own terms and not grounded in any broader conceptual or
philosophical systems. The German 'hermeneutics' school[24] has been used
for similar purposes. More recently, post-structuralist and postmodern
theorists have provided additional arguments for the insularity of Christian
discourse, as in Lyotard's argument that we should be suspicious of any
'meta-narratives'.[25] In the United States, the neo-pragmatism of Richard
Rorty has been used to similar effect.[26]

The dominant school in analytic philosophy of religion today has
provided sophisticated epistemological arguments to somewhat the same
end over the last thirty years. The first stage came when first-rate philosophers such as John Hick and Alvin Plantinga brought a distinctively
Christian voice to debates in the philosophy of religion, providing
reformulations of the classic arguments for the existence of God and
replies to common objections to Christianity such as the problem of
evil. Interestingly, this group of allies was first brought together by the
well-known attacks on the meaningfulness of religious language, for
instance the famous 'university debate' in which Anthony Flew used
John Wisdom's anecdote of the undetectable gardner to challenge
whether religious language could be cognitively meaningful at all. (The
critique was a sort of application instance of the general programme
of logical positivism, at least in its more explicitly anti-religious forms
such as A. J. Ayer's *Language, Truth and Logic*.) Christian philosophers and
theologians produced numerous essays and anthologies, interest grew,
and gradually their work began to constitute an identifiable research
programme in Christian philosophy.

In a second stage a group of philosophers, several of them from Calvin
College, began to challenge the *onus* that had been placed on Christian
thinkers to prove or provide evidence for their religious beliefs. Their
answer was a 'new' theory of knowledge, which they referred to, at first
jokingly and later seriously, as Reformed Epistemology.[27] Nicholas Wolterstorff attacked what they called 'evidentialism', the view that only those
beliefs should be considered justified for which the believer possesses
adequate evidence. In a complementary essay, Plantinga defended 'basic
beliefs' as fundamental to the structure of knowledge. Beliefs that are
truly or 'properly basic' are formed immediately by our belief-forming
mechanisms, such as the belief that I see a tree or had cereal for breakfast. The individual agent does have an obligation, when faced with a
potential 'defeater' – that is, a reason to doubt the veracity of her belief
– to come up with a 'defeater defeater', a reason to discount the criticism

and thus remain with her original belief; and if she cannot defeat the defeater, she should abandon her first belief. But she does not have the obligation to provide foundations for her religious beliefs in advance. Plantinga's contention was that 'God loves me' could be a basic belief just as much as 'I see a tree before me' could be. Of course, it follows directly from the truth of 'God loves me' that God exists; hence the Christian is fully justified in believing that God exists.

Since 1983 this movement in the philosophy of religion has increased in sophistication, in influence and in pages published. Wolterstorff has extended the argument for this theory of knowledge backwards to the beginning of modern thought and has begun to spell out a philosophical theology congruent with it.[28] William Alston has developed the analogy between visual perception and religious perception in great detail.[29] The philosophers in question have established a professional organisation (the Society of Christian Philosophers) and a journal (*Faith and Philosophy*), and have gained dominance within the field of philosophy of religion at large and at several key philosophy departments, most notably the University of Notre Dame (referred to, half-jokingly, as 'the philosophy department greater than which none greater can be conceived'). In recent years, building on the epistemological basis just sketched, they have moved increasingly into philosophical theology, exploring explicitly Christian intuitions and beliefs using the tools of analytic philosophy.[30] With the introduction of explicitly Christian assumptions and the vehement insistence that it is not necessary to refute opposing positions regarding the existence of God (e.g. atheism), the dividing line between the philosophy of religion, philosophical theology and even systematic theology has begun to disappear. A number of the publications of philosophers in this school are now indistinguishable from essays in systematic theology, although some show little knowledge of (or even interest in) the history of the theological tradition.[31]

An Argument for Accepting a Burden of Proof

What all the thinkers covered so far in this section share in common is their challenge to the claim that the Christian theologian or philosopher faces any sort of burden of proof in doing Christian theology or making Christian truth claims. I have argued in an earlier publication that no such strict dividing line can be drawn between the 'inside' and the 'outside' of Christian theology. A language whose meaning stems totally from Christian assumptions and the context of Christian practice does not exist.[32] The vocabulary theologians use has general meanings as well as specifically Christian meanings, Christian practices are derived from or show similarity with non-Christian practices, and Christians form their beliefs about the world and about how to act from sources other than Christianity – including science, early upbringing and enculturation, and widely shared experiences common to all or most human beings. Structures of plausibility cannot therefore be isolated from such broader contexts.

Take, for example, the question of religious pluralism. Plantinga has argued[33] that the existence of alternative religious belief systems in no way affects the justification of the Christian believer in holding her religious beliefs. In opposition to this view, Mark McLeod has recently shown that the existence of such alternative systems of belief, be they religious or secular, does affect the epistemic status of an individual's beliefs.[34] In a frontal attack on the entire recent trend in the philosophy of religion, Keith Parsons has further argued that this mistake about the burden of proof reveals a fundamental flaw in the work of Alvin Plantinga and Richard Swinburne. He concludes:

> The most important question for the philosophy of religion is not whether belief in God is rational, but whether that belief is true. As we saw, progress on this latter issue cannot be expected unless theists and atheists are willing to seek out a common universe of discourse, with shared canons of rationality and premises accepted by both sides.[35]

Preserving a broader discourse should be of interest to the religious thinker as much as to the atheist. Parsons tries to show that, of the two parties, it is the theist who has greater reason to take on the burden of proof, since she is the one who is advocating a particular belief, and 'those who lack a belief have no responsibility to defend that lack unless there is some reason that they should believe' (p. 147). The result seems to be a defeat for the theist:

> The longer theists are unable to meet [the] burden of proof, the less interested atheists will be in arguing about it. Atheistic apathy is likely to be encouraged when it is noted that Alvin Plantinga . . . expends vast labors of logic to prove that theism, at best, can only claim to break even with atheism. (p. 147)

Parsons likewise attacks Richard Swinburne's arguments for the existence of God, focusing in particular on his cosmological argument. Swinburne's claim, when one looks at it closely, is rather modest; he asserts only that the existence of the universe provides some support for the claim that God exists, although 'not necessary enough support to make it more probable than not that God exists'.[36] Parsons challenges Swinburne's positive case on a number of fronts. He questions whether probabilities can even be assigned to such abstract claims as 'the universe exists' (a reservation incidentally shared by Plantinga as well). He challenges Swinburne's assumption that simplicity is the only criterion for weighing the intrinsic probability of ultimate hypotheses (p. 78), from which it would follow that theistic explanations are massively more probable than is the brute existence of the universe in all its complexity. In an extended section (pp. 81ff.), he also rejects Swinburne's claim that theistic explanations are highly analogous to scientific arguments, showing (rightly, in my view) that significant disanalogies between scientific and theistic explanations exist, so that an identification of the two types of explanation appears mistaken. Citing Paul Davies, Parsons also questions whether the hypothesis of an infinite God is in fact more simple than a universe issuing from 'the simplest state of all', thermodynamic equilibrium, 'a state of infinite temperature, infinite density, and infinite energy'.[37]

Parsons begins his critique of Plantinga with a major concession, granting that theism may be a rational belief for many people: 'Atheists can admit without hesitation that religious experience is coherent, persistent, and, for many, compelling. Persons who believe in God on the basis of such experiences can therefore be regarded by atheists as perfectly rational' (p. 36). Still, Parsons insists, one can *defend* the existence of God 'only if arguments or evidence can be offered in support of that proposition.' It is this need for arguments or evidence that constitutes a certain, albeit limited, burden of proof (p. 38).

Theologians should note that both sides in this debate seem to have reached common agreement on several key points. Both agree, for example, that theologians are free to pursue theology in a systematic manner without having to place all of their work in an apologetic context or to provide arguments for each of their points. Atheists such as Parsons and the authors he cites (including Anthony Flew) also grant the rationality of religious belief for religious believers. Parsons wants to claim only 'that the denial of God's existence is, at least for some people, a perfectly rational and justified belief' (p. 38) – a claim that, I suggest, the Christian theologian need not dispute.

Admittedly, Parsons seems at times to miss the point of Plantinga's arguments by seeking to distinguish sharply between *the rationality of believers* and *the grounds for theistic belief*. It is Plantinga's contention that believers are rational because they have grounds for their belief – the same sort of experiential grounds that you might have for believing that there is a tree outside your window or that you ate cereal for breakfast this morning. What is at issue is the intersubjective nature of these grounds. In this sense, I suggest, the debate over 'foundationalism', as it is often raised by both sides, is a red herring. None of the parties in this debate really advocates self-evident intuitions in the sense of the great rationalist tradition of modern thought (Spinoza, Leibniz). Indeed, *both* sides seem to grant that one's individual experience can justify one in holding beliefs; whether or not one finally calls these beliefs 'basic beliefs' is a matter of terminology. Idiosyncratic factors, whether purely personal or related to a group or culture, clearly play some role in what the individual is justified in believing.

The question is instead: under what conditions ought an individual (or group) either (1) to question her own beliefs based on arguments or disagreement from others, or (2) to accept some onus to provide reasons for her beliefs that might be persuasive to those outside her particular group? Theologically, one might say that Plantinga is the Barth (indeed, more Barth than Calvin) of the contemporary debate in the philosophy of religion. According to him, it does not matter at all that there are competing religious beliefs; it does not matter that non-Christian scholars come to very different conclusions about the historical veracity of the New Testament documents; it does not matter that most biologists come to conclusions about the development of life very different from Plantinga's conclusions. Admittedly, Plantinga does insist that the believer comes under some obligation when she is confronted with potential 'defeaters' for her beliefs. But what constitutes a potential defeater turns out to be a highly controversial and somewhat subjective phenomenon, thereby reducing the onus that one might have expected. For example, the fact that most of the scientific establishment might disagree seems, for example, not to represent a worrisome defeater for the Plantingian biologist. Here I agree with Parsons:

> Suppose that the dictator of a small country begins to call himself 'The Almighty' and insists that all citizens prostrate themselves before him and address him only in the most worshipful tones. Suppose further that he declares his divinity to be a self-evident truth and orders that all doubters be put to death as blasphemers. Surely *something* is seriously wrong with the dictator's claims to divinity; surely *that* cannot be a properly basic belief. The same judgement would seem to hold with respect to such beliefs as 'Killing innocent people for fun is morally permissible' and 'I can

fly simply by flapping my arms'. To allow such beliefs to count as properly basic seems tantamount to having no standards of rationality at all. (p. 43)

Now these are moral examples, involving the treatment of other humans, whereas religious beliefs seem at first glance to be much more private. But 'private' religious beliefs often have broader social significance, if only because of the attitudes which they sometimes engender. One need think only of the actions, ranging from prejudice to outright terrorism, of, for example, Islamic fundamentalists. There are thus instances in which a privately formed belief, even one that might seem to be a basic belief to the individual agent, *ought* to be questioned because of its moral implications or its conflict with more universal human experience.

Conclusions

Plantinga, like Barth, does not want to see Christian beliefs subjected to any 'universal principles of human reason'; he does not want them to be deemed questionable *until* they meet such standards nor rejected because they conflict with them. I have conceded that the best current account of rational warrant agrees with Plantinga in disputing the evidentialist requirement as a necessary condition for holding beliefs rationally. At the same time, it is the thesis of this book that (1) in cases of outright conflict between Christian knowledge claims and our best current knowledge of the world, a clear burden of proof rests on the Christian to address the conflict or alter her beliefs, and (2) in other cases it is recommended (or highly desirable) to solicit and to respond to broader feedback of this sort. Some of the reasons for accepting this requirement are

1. Believers are themselves rational agents who have internalised structures of plausibility from their education and upbringing. To ask them to ignore conflicts is to create a schism and engender a sort of epistemic schizophrenia.[38]
2. To ignore challenges from science, philosophy or other religions, and especially to ignore challenges of moral questionableness in Christian truth claims, is to foster the impression of not caring – or, worse, of the theologian's inability to respond to the charges.
3. The main reason, however, is theological: if God is the source of all truth, then there should be a consonance between the conclusions of human scholarship in general and the theological conclusions based upon revelation.

4. The impression should not be raised that we can, using some sort of natural theology, argue people into the kingdom of God. Nonetheless, reasons can often be given to show the plausibility, the credibility, of Christian claims.

5. The process of coming to understand revelation more fully entails an interaction with everything that the believer knows. This is the classic principle of *fides quaerens intellectum*, or faith seeking understanding – the task of integrating together all that one knows, which (viewed theologically) is part of the ongoing process of sanctification.

6. Some apologetic work is mandated by the biblical texts themselves, e.g. the exhortation to 'be prepared at all times to give a defence (*apologia*) for the faith that is within you' (1 Pet. 3: 15).

7. It seems further that theologians sometimes *do* in fact have the better arguments – that, for example, a theistic understanding of the origin of life constitutes a better explanation of the phenomena than does the explanation of 'unguided evolution' *à la* Darwin.[39]

8. To demand insulation from critique would require the Christian theologian to extend the same 'tolerance' to other religious positions. But in so far as she holds that Jesus Christ is the Son of God, she may well have some interest in critically examining the arguments of others – not to mention exposing her own arguments to criticism. But this is possible only if religious positions are not isolated from one another.[40]

9. What constitutes a 'properly basic belief' is somewhat ambiguous; the line will be, of necessity, difficult to draw. Thus we should, whenever possible, provide a broader justification for beliefs that may be for us individually basic.

These are the reasons that compel me in this text to attempt to bring theology into closer dialogue with the sciences and philosophy – not because the theologian faces a burden of proof which makes it irrational for her to hold Christian beliefs apart from apologetics, but because theologians can, and in at least some cases should, interact with what humans seem to know about the world in the attempt to provide the best possible explanation, from a Christian perspective, of those data. To engage in this task *just is* to do theology 'in an age of science'.

Through our initial study of the biblical creation stories and the doctrine of God's relation to the world in Christ we have established the parameters for a Christian account of God and God's action in the world. It should now be possible to proceed to the task of integrating this theological account with the best that we know about the world through science and philosophy.

NOTES

1. See Hans W. Frei, *The Eclipse of Biblical Narrative: A Study in Eighteenth and Nineteenth Century Hermeneutics* (New Haven, CT: Yale University Press, 1974); *The Identity of Jesus Christ: The Hermeneutical Bases of Dogmatic Theology* (Philadelphia: Fortress Press, 1975); *Types of Christian Theology*, eds George Hunsinger and William C. Placher (New Haven, CT: Yale University Press, 1992); *Theology and Narrative: Selected Essays*, eds George Hunsinger and William C. Placher (New York: Oxford University Press, 1993); and Garrett Green (ed.), *Scriptural Authority and Narrative Interpretation* (Philadelphia: Fortress Press, 1987).
2. See Frei, *The Identity of Jesus Christ*.
3. See Hans-Georg Gadamer, *Truth and Method* (New York: Seabury Press, 1975).
4. In the new Systematic Theology Group of the American Academy of Religion over the last four years we have observed such rethinking and reappropriating going on. Seeking to mediate between scholars who tell us what the doctrines originally *meant* and those who speak of what *needs to be said* in today's context, a number of theologians have attempted to think these beliefs afresh, to ask anew about their meaning and significance in both traditional and radically new terms. It is hoped that the Edinburgh Series in Constructive Theology might serve a similar function in a broader context.
5. This is the classic definition of God developed by Rudolf Otto in his *The Idea of the Holy: An Inquiry Into the Non-Rational Factor in the Idea of the Divine and its Relation to the Rational*, trans. John W. Harvey (London: Oxford University Press, 1923).
6. See Joseph Bracken, *The Divine Matrix: Creativity as Link Between East and West* (Maryknoll, NY: Orbis Books, 1995). I owe much to Prof Bracken's theological reflection, which draws (as I do) from classical theological (in his case, Thomistic) reflection, the categories of German Idealism, and the insights of twentieth-century process thought, particularly Alfred North Whitehead. See, for example, his recent collection on classical and process trinitarianism, co-edited with Marjorie H. Suchocki, *Trinity in Process: A Relational Theology of God* (New York: Continuum, 1997).
7. For Pannenberg on the Trinity, see especially his *Systematic Theology*, trans. Geoffrey W. Bromiley (Grand Rapids, MI: Eerdmans, 1991), Vol. 1, Chapters 5–6.
8. See especially Falk Wagner's philosophical (Hegelian) doctrine of the Trinity in Wagner and Friedrich Wilhelm Graf (eds), *Die Flucht in den Begriff: Materialien zu Hegels Religionsphilosophie* (Stuttgart: Klett-Cotta, 1982).
9. Karl Barth, *Church Dogmatics*, ed. and trans. G. W. Bromiley and T. F. Torrance (Edinburgh: T. & T. Clark, 1956), II/1, para. 30.

10. Barth, *Church Dogmatics*, II/1, pp. 598f.
11. See Wolfhart Pannenberg, *Theology and the Philosophy of Science*, trans. Francis McDonagh (Philadelphia: Westminster Press, 1976). See also Philip Clayton, *Explanation from Physics to Theology: An Essay in Rationality and Religion* (New Haven, CT: Yale University Press, 1989), and Nancey Murphy, *Theology in the Age of Scientific Reasoning* (Ithaca, NY: Cornell University Press, 1990).
12. It would also be desirable to write a 'theology of nature in light of the world religions', a work that is urgently needed if we are to have the full resources for reflecting on the theology of nature today.
13. Barth, *Church Dogmatics*, III/1, p. ix.
14. Admittedly, it might well have been necessary in his day, to help break theology from 'natural theology' and from the liberal theology of Adolf von Harnack and others. Our question is whether the Barthian limitation remains necessary and prescriptive for theology today.
15. For a much more detailed treatment of the anthropological themes see Pannenberg's *Anthropology in Theological Perspective*, trans. Matthew J. O'Connell (Philadelphia: Westminster Press, 1985).
16. This is also a point on which Pannenberg has insisted. See Clayton, 'The God of history and the presence of the future', *Journal of Religion* 66 (1986): 95–105.
17. Barth, *Church Dogmatics.*, III/1, p. 81.
18. See John Hick's publications since the early 1970s, and most especially *An Interpretation of Religion* (New Haven, CT: Yale University Press, 1989).
19. 'Inside' and 'outside' are relative terms; I have argued elsewhere that there is no *absolute* inside or outside (Clayton, *Explanation*, Chapter 5).
20. For a good introduction to the field, see Michael Peterson, William Hasker, Bruce Reichenbach and David Basinger (eds), *Reason and Religious Belief: An Introduction to the Philosophy of Religion* (New York: Oxford University Press, 1991).
21. Hans Frei, *The Eclipse of Biblical Narrative: A Study in Eighteenth and Nineteenth Century Hermeneutics* (New Haven, CT: Yale University Press, 1974).
22. George Lindbeck, *The Nature of Doctrine: Religion and Theology in a Postliberal Age* (Philadelphia: Westminster Press, 1984); see especially the Excursus to Chapter 3.
23. In addition to Lindbeck, D. Z. Phillips has made this case in numerous publications. See, for example, *Faith After Foundationalism* (London: Routledge, 1988).
24. See especially Gadamer, *Truth and Method.* Among many secondary works see Josef Bleicher, *Hermeneutics as Method, Philosophy and Critique* (London: Routledge, 1980).
25. See Jean-François Lyotard, *The Postmodern Condition: An Essay on Knowledge*, trans. Geoff Bennington and Brian Massumi (Minneapolis: University of Minnesota Press, 1984).
26. See Richard Rorty, *Contingency, Irony, and Solidarity* (New York: Cambridge University Press, 1989). For a theological application of Rorty's 'neo-pragmatism', see the essay by C. Wesley Robbins in *Zygon* 28 (1993), and

the responses by Clayton and Wenzel van Huyssteen that follow in the same volume.

27. See Alvin Plantinga and Nicholas Wolterstorff, *Faith and Rationality* (Notre Dame, IN: University of Notre Dame Press, 1983).

28. See Nicholas Wolterstorff, *John Locke and the Ethics of Belief* (Cambridge: Cambridge University Press, 1996) and *Divine Discourse: Philosophical Reflections on the Claim that God Speaks* (Cambridge: Cambridge University Press, 1995).

29. William Alston, *Perceiving God* (Ithaca: Cornell University Press, 1994).

30. See especially the works of Thomas Morris, e.g. *Anselmian Explorations* (Notre Dame, IN: University of Notre Dame Press, 1983) and *The Concept of God* (Notre Dame, IN: University of Notre Dame Press, 1989).

31. An interesting comparison of the methods and assumptions of analytic philosophers of religion and theologians within the American Academy of Religion has just been published: William J. Wainwright (ed.), *God, Philosophy and Academic Culture: A Discussion between Scholars in the AAR and the APA* (Alpharetta, GA: Scholars Press, 1996).

32. See Clayton, *Explanation*, especially Chapter 5, e.g. pp. 132ff.

33. In an address presented at the Pacific Division of the American Philosophical Association in Portland (1993).

34. See Mark McLeod, *Rationality and Theistic Belief: An Essay on Reformed Epistemology* (Ithaca, NY: Cornell University Press, 1994).

35. See Keith M. Parsons, *God and the Burden of Proof: Plantinga, Swinburne and the Analytic Defense of Theism* (Buffalo, NY: Prometheus Books, 1989), p. 146.

36. Parsons, *God and the Burden of Proof*, p. 72.

37. Parsons, *God and the Burden of Proof*, p. 95. Parsons' attack on Swinburne's criterion of simplicity is less convincing. It is widely accepted that 'aesthetic' criteria such as symmetry, simplicity and 'fit' or coherence are fundamental to the pursuit of theoretical physics today. This is yet another case where physical theory shades over gradually into metaphysical criteria. In attacking this criterion Parsons seems to be relying on an older, empiricist model of science that does not correctly express the way that cosmologists and theoretical physicists work today. Thus the simplicity of theistic explanations should continue to recommend them to us rationally, even if Swinburne is wrong in employing theological explanations as if they were identical to scientific explanations.

38. See my notion of the 'secular believer' in *Explanation*, Chapter 5.

39. See my 'Inference to the best explanation', *Zygon* 32 (1997, 377–91).

40. 'After all, what would Plantinga say to a Shirley MacLaine who claims to recall her past lives, or to a Jeane Dixon who claims to receive guidance from the stars, or, indeed, to a Linus Van Pelt who is sure that his pumpkin patch is the most sincere and will be visited by the Great Pumpkin this Halloween?' (Parsons, *God and the Burden of Proof*, p. 50).

4

RETHINKING THE RELATION
OF GOD AND WORLD:
PANENTHEISM AND THE
CONTRIBUTION OF PHILOSOPHY

———·◦◦◦◦◦◦◦◦——

We have begun with the biblical data and devoted two chapters to them. Obviously, scientific and philosophical questions could not be totally suspended during this inquiry. To pretend otherwise would have been to have fallen back into a form of 'theological positivism', one which claims that there is a privileged route, accessible in principle to all reasoners, from the biblical data to the (objectively speaking) best systematic theology, the best theory of God. But the biblical narratives, like those of other traditions, *cannot* be fully captured in a context- and time-independent fashion. This book is premised on the assumption that there is a hermeneutical rather than a linear relationship between biblical interpretation and theological reflection. It is no more possible for one to exclude from the interpretive process what one finds scientifically and philosophically (and ethically!) credible than it is for one to read individual texts about God's creating and redeeming without being influenced in some way by one's broader theologies of creation and redemption. There is no shortcutting the hermeneutical circle.

One can nonetheless shift its centre of focus. We will find (in Chapter 5 below) that productive discussion between theology and the sciences requires finding some third playing field within which the similarities and differences between their two sets of conclusions can be brought to

82

clear expression. Before we turn our primary attention to the doctrine
of God's activity in the world in light of contemporary science, then, it
behoves us first to explore what kind of a contribution philosophical
reflection can make to the doctrine of God. This chapter's question, then,
is: what general framework for conceiving of the God/world relation is
adequate both to the biblical data and to the contemporary philosophical
and scientific context?

MOVING FROM THE TEXTS TO THE
DOCTRINE OF GOD

It will be my thesis in the coming pages that a framework has been
developed in relatively recent times which offers new prospects for
specifying the complex dimensions of the relationship between God and
world in a way that is both conceptually adequate and does justice to
the biblical narratives. If this hypothesis is correct, it will provide some
new resources for thinking about divine activity in the world even in
light of the best that recent science has to offer.

From Polytheism to Monotheism

The philosophical position that was the natural ally to Christianity in its
early centuries I shall call *classical theism*. In the most general sense, theism
means simply the belief in the existence of a God or gods – one or more
supernatural, personal beings. Within theism, we distinguish between
monotheism, or the belief in the existence of one God, and *polytheism*, or
the belief in many gods. Not monotheism as much as Greek polytheism
was the dominant view that early Christian thinkers found around them
as they began to reflect conceptually on what it was they had come to
believe. As is well known, Greek polytheism held that there were
multiple gods who exercised a causal influence over specific events in
the world. The analogies between the gods and the actions of human
agents were quite pronounced in Greek polytheism: the gods played
favourites among humans, quarrelled constantly and seemed preoccupied
with petty concerns such as the rivalries with each other and between
their heroes on earth.[1] One often has the sense that, were it not for their
greater power and the fact that they were immortal, the gods would be
indistinguishable from Greek men and women of that time.

In polytheism there is no 'problem of divine agency'. It was simply
taken for granted by the Greeks, as it was by believers in Babylonian,
Egyptian and Indonesian polytheisms, that the gods exercised a high (if

not fully consistent) level of power and ability to influence events in the world. They not only *could* alter human affairs, they did so with a high degree of frequency.

Since most cultures have felt that the origin of the world needs to be accounted for, and since gods are obviously more natural candidates than human agents for explaining the world as a whole, in most polytheistic religions one or more of the gods are given responsibility for the origin of the world. This use of the gods in *cosmogenies* (theories of the birth of the world) should actually have led the Greeks and others to rethink the gods' relationship to the world, for reasons we will explore in detail. But the logic of the creation of the world was not fully thought through at that time, and the full implications of a creation of the world were not drawn. One of the reasons that creation did not pose the puzzle it should have posed was that matter – that out of which the world is made – was taken to be pre-existent. This suggested that creation was perhaps not *all that different* from the everyday acts in which human agents mould and form pre-existent things.

Obviously, making the radical switch from many gods to belief in one god would change the relation of this god to the world. A single god would no longer be one member of a group, limited by the others and describable in terms of their common characteristics. Instead, every quality ascribed would be ascribed to him or her alone. It would be difficult, for example, to separate between the god's *contingent* features and what pertained to him or her essentially, as an inherent part of his or her nature. Clearly the Greek or tribal peoples' gods were limited in power, since they could not destroy one another; but what would limit the divine power if there were only *one* god? Should this god be held to be the origin of *matter* as well, as in the doctrine of *creatio ex nihilo*? How would this god's power and relation to the world *then* need to be understood? Clearly, in this case we are speaking of a very different kind of belief claim – a being so radically different from a member of a panoply of gods that the transformation appears qualitative rather than quantitative. It actually appears that the resultant theistic belief involves *a different kind* of being rather than merely the same kind of being but with lots more power.

The Hebrew writers were not totally blind to these changes. From the very beginning they encountered a god who was 'jealous' of any competitors ('for I, Yahweh your God, am a jealous God, visiting the iniquity of the fathers on the children, on the third and fourth generations of those who hate Me', Exod. 20: 5). Allegiance and obedience to Yahweh had to come above all other concerns; there could be 'no other gods' before him (Deut. 5: 7). Yet the allegiance would pay off, the Israelites found, for this God also *proved* himself more powerful than all the others; he demanded the highest position *because* he had a power unrivalled by the others. The assumed context in the earlier Hebrew texts is polytheism: 'For all the peoples walk each in the name of its god, but we will walk in the name of Yahweh our God for ever and ever' (Mic. 4: 5). Likewise, the Psalmist praises Yahweh because he 'is a great God, and a great King above all gods' (Ps. 95: 3).

Radical Monotheism

Because of the dissimilarity in power and the moral demands raised by Yahweh, the Yahwistic texts then gradually began to speak of God as the 'only true God'; all others were in some sense imposters. Gradually – and this may be the most important development of all – *the Israelites began to speak of Yahweh as the only God*. If Yahweh really was who he claimed to be, there simply could not exist any other gods beside him; he countenances no rivals, they came to understand, because ultimately he *has* no rivals. One finds in the development of the Yahweh cult in the Hebrew Bible a fascinating narrative of precisely this development. The most ancient strata of the tradition still contain references to the other gods, as in Psalm 82: 1 ('God has taken his place in the divine council; in the midst of the gods he holds judgement'). But in later strata there *are* no longer any competitors to Yahweh: 'Turn to me and be saved, all the ends of the earth! For I am God, and there is no other' (Isa. 45: 22).

The implications of this metamorphosis for the Israelite theological tradition (and those to which it gave birth) cannot be overstated. Yahweh had to be approached in a completely different fashion to the gods of the surrounding peoples. The 'other gods' were first weaker, and then 'false gods', and finally *no gods at all*. Gradually the Israelites realised that, in so far as Yahweh's power was not limited by other divine agents, he would be immensely more powerful than any polytheistic concept of 'the highest god'. Except perhaps for the resistant properties of matter – which might refuse to be moulded fully according to God's will but might instead resist it – nothing could stand in the way of God actualising his own purposes. And should it be that the earth was 'without form and void' in the beginning because *nothing* existed before Yahweh created, then in this case Yahweh's creative powers, his ability to choose and his responsibility for what was created, will be truly infinite.

The same line of thinking that led the Israelites to realise that there

85

could be no other gods alongside Yahweh finally also led Jewish and Christian theologians to claim that God's creative activity could not be limited by some pre-existent matter, binding his hands and limiting his creative choices. *Monotheism*, after all, means the existence of one and only one god, in contradistinction to *henotheism*, which involves belief in a single god *without* the assertion that there is only one god. Eventually it became clear that the logic of monotheism actually required that God be understood as the origin of matter as well. God's power would not be absolutely unlimited if a foreign substance, matter, existed apart from him and could therefore not be explained by him. Thus the doctrine of *creatio ex nihilo* was born: God, whose power was absolutely unlimited, existed as the only entity or 'thing' in the beginning. When he created the world, he also created the matter out of which it was composed.

This is a very 'high' doctrine of creation – it is difficult to imagine God's power being any greater. The trouble is, aside from the initial creation, remnants of polytheism remained in the picture of God and God's actions – or, to put it more carefully, the theological tradition continued to retain elements of henotheism. Certain theological formulations continued to conceive of God as if he were one among many rather than understanding God as the absolutely unique and only divine Source. Recall that the gods of polytheism frequently stand outside the world (e.g. on a heavenly mountain) and influence it from where they stand – sometimes entering into the world to take a lover or assist a warrior, as in the Greek myths, sometimes intervening at crucial moments (e.g. the warrior's rite of passage), sometimes transmitting their power through intermediaries. The Jewish and Christian traditions have not been immune from a henotheistic way of thinking; they have sometimes failed to be radical enough. Although they obviously moved from many gods to one God, transforming the notion of creation appropriately, they have often continued to conceive of God as *a being* who stands alongside the world, which becomes a 'handiwork' he has crafted. The picture of traditional monotheism, we might say, is of a God who existed before the world and who found a place near himself in which to create the universe. This picture might indeed give rise to that often reproduced medieval sketch, which shows the explorer climbing past the last heavenly sphere and opening a sort of doorway in the heavens, through which he looks outward into the realm in which God dwells.

It is my thesis that this mixture of pictures and metaphors suggests that a significant part of the theological tradition has not thought through the implications of strict monotheism carefully enough. Further, careful historical study shows that it was no coincidence that this particular blend – a monotheistic doctrine of creation and a view of divine agency reminiscent of henotheism – played such a central role within the history of theology. For it appeared at the time that the only alternative was *pantheism*, the view that 'all things *are* God'. Yet pantheism obviously flew in the face of the biblical documents, which insist on an ontological distinction between God and the things he created: genetically, God is the source of all, and morally he sets absolute moral standards for his followers. In philosophical terms, the biblical God is infinite and all creation is finite; God is necessary and all other things are contingent. Classical Western theism or pantheism – if this is the choice, then clearly the Christian theologian must choose classical theism.

H. Richard Niebuhr has provided one of the classic treatments that distinguishes partial from 'radical' monotheism. Niebuhr defines *radical monotheism* as follows:

> For radical monotheism the value-center is neither closed society nor the principle of such a society but the principle of being itself; its reference is to no one reality among the many but to One beyond all the many, whence all the many derive their being, and by participation in which they exist. As faith, it is reliance on the source of all being for the significance of the self and of all that exists. It is the assurance that because I am, I am valued, and because you are you are beloved, and because whatever is has being, therefore it is worthy of love. It is the confidence that whatever is, is good, because it exists as one thing among the many which all have their origin and their being, in the One – the principle of being which is also the principle of value. In Him we live and move and have our being not only as existent but as worthy of existence and worthy in existence. It is not a relation to any finite, natural or supernatural, value-center that confers value on self and some of its companions in being, it is a value relation to the One to whom all being is related . . . Radical monotheism dethrones all absolutes short of the principle of being itself. At the same time it reverences every relative existent. Its two great mottoes are: I am the Lord thy God; thou shalt have no other gods before me and Whatever is, is good.[2]

Niebuhr recognises (pp. 34f.) that radical monotheism requires broader loyalty than loyalty to one group; indeed, as he points out, it transcends even loyalty to humanity. Thus, for example, radical monotheism clearly implies a more extensive environmental ethic than Christianity has developed in the past. Indeed, radical monotheism might even be a *necessary condition* for a fully adequate environmental ethic, since it so strongly maintains that humankind is no longer the centre of being and value. (I believe that the same strengths are found in pantheism in both its classical Eastern forms and in some of its Western manifestations, such as recent appropriations of Spinozism within environmental ethics.[3])

Now some aspects of Niebuhr's conclusion should give us pause; the principle of being, for example ('Whatever is, is good'), may need the same sort of relativising as the other principles.[4] But Niebuhr's work does show with particular clarity why we need a more radical conception of God. Biblical ideas about God give rise to a process of systematic reflection that lifts one beyond individual stories of God's action in the world, beyond doctrines of God as the most powerful among the worldly forces, to an understanding of God as absolute origin, as the power that holds all finite things in being, as Goodness itself. Could it be that this movement pushes us yet one step further, beyond radical monotheism as Niebuhr conceives it – and yet not in the direction of pantheism?

FROM CLASSICAL THEISM TO PANENTHEISM

It turns out that 'classical Western theism or pantheism' is *not* the only choice. One sign of the inadequacy of this dichotomy – despite its rhetorical role within the history of theology – is that pantheism was never seriously entertained within the orbit of Christian or para-Christian reflection. Always – in Porphyrus, in Johannes of Eriugena, in Eckhardt, Cusa, Bruno and many others – the actual 'heretical' assertion was that finite things are *in* God; the world as a whole was never made identical to all there was of God. Admittedly, as the church struggled over the question of which of these views to label heretical, the debate was indeed carried out in terms of whether the positions were pantheistic or not. But this was to turn a complex conceptual struggle into a matter of superficial labels, to attempt to dismiss a sophisticated theological position by means of the fallacy of 'guilt by association'. In what follows I hope to show that panentheism dissolves the dichotomy that structured so many of the theological debates on this topic.

The Argument from Space

In *God in Creation* Jürgen Moltmann shows convincingly why a full understanding of creation must eventually lead theologians to something like panentheism, the position which he himself advocates, and not to pantheism in the sense that the defenders of orthodoxy feared. In

Moltmann's presentation what plays the key role is closer reflection on the question of *God's relationship to space and time*. On the one hand, the Christian tradition has never reduced the world to unreality or illusion. There really are separately existing things, which means that they are spread out in real space and time. As Moltmann notes, 'If space is interpreted as the dimension of God's omnipresence, pantheistic conclusions are impossible.'[5] God can only be present to all parts of his creation if there really are such parts in the first place.

On the other hand, space and time must be thought theologically, so that their origin within God becomes clear. For example, it is not enough to claim, following Leibniz, that each object or 'monad' gives rise to its own space and time. Yet neither can space and time be independently existing 'things', such that, before creation, there existed a sort of huge spatio-temporal box, totally empty, which God then 'filled' with a variety of stuff when he created.[6] More adequate is a view closer to Newton's own, according to which space is a sort of divine *sensorium*, a framework imposed by God on what he perceives in so far as he perceives it as not identical with himself. Of course, Newton's view must be corrected by the insights of relativistic physics, since we now know that the space 'within' which objects exist is affected or 'curved' by those objects themselves. Yet Newton did see that, theologically, space must be understood also as an attribute of God, and hence as part of God.[7]

One cannot help but note the analogy with Augustine's theory of time. Augustine allowed the entire flow of time to be encompassed within the eternal Now, which expresses God's awareness of all temporal moments at once. As God can be present to every now while still subsuming all now's within the eternal Now that transcends and encompasses finite time, so also God can be present here while still subsuming all here's within a divine space that transcends and encompasses physical space. (The analogy is strengthened by the role of 'spacetime' in contemporary physics.) Thus Moltmann can conclude, ' "Absolute space" means the direct presence of God in the whole material world and in every individual thing within it.'[8] The motivation of the doctrine of divine omnipresence, then, is not to pretend that all things *are* God, but to locate all things *within* the divine presence, which is the only source of all existing things. If space is an attribute of God, then God must be present at all points in space. Theologically, 'if God perceives everything immediately and directly through his omnipresence, this presupposes that God's eternal, uncreated omnipresence is *the same as* the omnipresence of space'.[9] If space is God's space, then the world is not 'outside' him but by definition within him.

It thus seems that we must not separate the world's 'space' from God's omnipresence. One might worry that pantheism will result, since space (and thus all finite things) are no longer placed outside of God. But pantheism means that *no* separation is made between God and the world, whereas I am suggesting that the separation ought to be drawn in a different way. Fear of pantheism drove theologians to use spatial difference as the 'specific difference' between God and world when they should have trusted the power of more fundamental theological categories: finite versus infinite, contingent versus necessary, imperfect versus perfect – created versus Creator. We are not God because we are different *in our fundamental nature* from God. Thus it does not matter where we are located: within the overarching divine presence, and even (in one sense) within the divine being itself, we remain God's created product, the work of his hands.

A word should be said here about the finite/infinite distinction, since it plays a very important role in panentheism. It is never merely a quantitative distinction, as if the infinite were merely a bit more than the finite – or even *a lot* more! Like the difference between contingent and necessary, it expresses a qualitative distinction, a difference in nature or being. This point is especially crucial because finite space *might* turn out to be endlessly extended and there might be an uncountable number of objects (say, atoms) within it. Nonetheless, as Georg Cantor, the founder of the theory of transfinite sets, realised, even a mathematics of infinite sets still requires as its limit case the notion of the absolute infinite, which is qualitatively different from any infinite set.[10] Applied to space, this means that even an endless ('infinite') space could be included within God without being identified with him. To preserve Cantor's distinction theologically, we might say that God encompasses infinite (created) space but that God *is* absolute space. This distinction makes it possible to think of God as coextensive with the world: all points of space are encompassed by God and are in this sense 'within' him. Nonetheless, created space is precisely that – created, contingent. Only God himself has the ontological status to be absolute and to contain all space within himself. In short: finite space is contained within absolute space, the world is contained within God; yet the world is not identical to God. Precisely this is the core thesis of panentheism.

Admittedly, this view calls for a dialectical way of thinking that has not always been embraced within the Christian tradition. Not infrequently, the God/world question has been posed in terms of an either/or: *either* the world is separate from God and exists 'outside of' him, *or* the world and God must be identified, in which case there can be no distinction between the two – and pantheism results. To think of the world as within God and at the same time as different from God is to think in terms of a both/and: there is *both* identification or inclusion *and* distinction of God and world. Of course, it would be conceptually 'cleaner' if one could specify in which respects the two are identified and in which others they are separate. For instance, why not just say that the world is in God spatially and yet distinct from God as the contingent is distinct from the necessary? Such formulations are helpful – and yet none of them, in my view, completely abolishes the tension.

In order to be theologically adequate, every assertion of the independent existence of the world must be balanced by the (equally true) assertion that the world is absolutely dependent upon God at every instant for its continued existence. The world really exists, *and* the world really depends on God moment to moment for its sustenance, such that nothing would exist in the next instant without the conserving will of God. Physical matter/energy really exists, *and* physical matter/energy has no independent existence apart from God. Such dialectical thinking is fully familiar to students of contemporary environmental science, where it appears in organic form in the *symbiotic* relationship between individual organisms and their life-world of other living beings and objects. We know that ecosystems are more than the sum total of living and non-living things within them. Now imagine an ecosystem of 'all that is'; it too must be more than the sum total of its parts. If you also imagine that its identity is living and conscious, and (for this is the claim of theism) that its existence also *preceded* its being filled with living things, then you will have some sense of the dialectic between God and world envisioned by panentheism. Even if the dialectic is not reducible, it remains conceptually consistent and theologically fruitful – as I trust the remaining chapters will show.

The Moral Difference

In the previous sections we moved from the doctrine of creation (Chapter 2) to the relationship between God and world that it expresses. We found various ways of maintaining the difference between God and world within the framework of a strong view of divine omnipresence, according to which all finite things exist within God. For instance, we found it possible still to retain the ontological difference between necessary and contingent, parallel to the distinction between infinite and finite. But the Christian tradition (like the Jewish and Muslim traditions) has also emphasised the moral difference – the difference of a God who is pure Goodness from human beings whose freedom is defined by their potential (or proclivity!) to do what is morally imperfect. Both as individuals and as a species, we manifest our ontological difference, our not-being-God. In fact, of course, the moral and the ontological dimensions are linked; as Westermann writes, 'In the biblical account of the origins, sin is not the narrow, individualistic notion that it has become in church tradition. It is viewed in a broader perspective. It is seen as that other limit, that inadequacy or overstepping of limits which determines the whole of human existence.'[11]

The ontological and moral questions are also closely connected, as the theological traditions early on realised, because the movement into sin represents the human wish 'to be God' and to act as if one were God. To say that humans are fallen is to say that they seek to create, and to possess, in the same manner that God does. The result is not only that God's place *vis-à-vis* creation is usurped; it is also that the unity of creation, in which all finite things are subordinated to their infinite source and thus take their place within the one overarching order of creation, is broken. In W. B. Yeats's famous lines:

> Things fall apart; the centre cannot hold;
> Mere anarchy is loosed upon the world,
> The blood-dimmed tide is loosed, and everywhere
> The ceremony of innocence is drowned . . .[12]

'Sin' means that the order of the whole is broken and fragmentation begins to manifest itself – the first appearance of that fragmentation which so fundamentally characterises the state of human existence at the end of the twentieth century.

The result of broken order, which the Christian tradition has understood as the attempt to usurp the place of God, is simultaneously ontological and ethical. After the fall, as in the story of Cain and Abel, man commits violence against man, man against woman, humans against each other, humanity against nature.[13] The step from a changed attitude toward the other to actual violence among the descendents of Cain and Abel is too obvious to require further comment.[14] I have often used the ontological pairs finite/infinite and contingent/necessary in characterising the world/God difference in panentheistic terms, but they need not exclude the moral difference as well. A metaphysics of perfection may face problems today that it did not recognise in the pre-modern period.[15] Still, the well-known problems do not prevent divine moral perfection from serving its two classical functions: distinguishing God from God's creation and helping to inspire ethical action.

Can the World Affect God?

To many religious persons who are theists, it may seem like a foregone conclusion that the world and its inhabitants can affect God. Does not Christian practice, for example, involve prayer, petition and praise to God for his guidance of history and individual human lives? Yet Western theology has struggled long and hard over the proposition that God is affected by the world. One can perhaps understand the motivations for this concern: when God is affected by an event in the world, he becomes different in some sense than he would have been otherwise. If this is true, it might seem that the world comes to play a necessary role in God's own history, since God would not have been the same without the world. Yet such a conclusion is unacceptable to traditional theologians, since creation must be the pure product of free divine choice and in no sense necessary.

As formulated, the objection rests on some modal mistakes, however. It could be that *if* God creates a world and is involved with its history in anything like the way the biblical documents suggest, then God will have certain properties at the end that he would not otherwise have had (say, the property of having given Teresa renewed hope at some time). But it does not follow that such acquired properties would change God in his essential nature. Also, there is no reason to think that God's initial decision to create thereby becomes necessary. A free creation remains free; any effect the world subsequently has on God is a consequence of the initial free decision rather than a sign of eternal necessity.

Nonetheless, even *if* conceptual arguments suggested some unwelcome consequences, I am not sure they could outweigh the overwhelming importance of the religious and the biblical arguments for the world's effect on God. The Hebrew Bible portrays a God who is intimately involved with his people, and the New Testament raises this involvement to the level of a parent's intense preoccupation with his or her child ('Abba, Father', Mark 14: 36; Rom. 8: 15; Gal. 4: 6). One can of course interpret every biblical passage that suggests a human influence on the plans of God as blatant anthropomorphism. In fact, however, the finger of suspicion points in both directions, for such reinterpretations of the texts may also reflect an overemphasis on 'philosophical' attributes such as immutability and aseity. Does not the outcome look very different when one brings to the text less the 'masculine' values of impassivity and self-sufficiency and more the 'feminine' values of responsiveness, community and inclusion?[16]

There may be limits on how far one can go in the direction of God's dependence on the world if one wishes still to be guided by the biblical texts. Still, the texts appear to suggest (even if the Western philosophical tradition before 1750 does not!) that the world can have an effect on God, even if not on his essential nature. From Hegel through Pannenberg, the central role of time and history has become increasingly clear to theologians.[17] Our own century has seen highly significant developments in conceiving God and process, beginning with the monumental work of A. N. Whitehead[18] and continuing, among many others, in the numerous works in which Charles Hartshorne challenged and moved beyond the immutability of classical theism.[19] Once one has challenged the hold on theology of a metaphysics of substance, which Christianity had inherited from Greek thought, and has seen how a metaphysics of process can provide just as adequate a conceptual framework for thinking about God, then one is hard pressed to say why God's perfection should require that he be immutable, unaffected by anything that goes on in the world. God can respond to the world's suffering, to the joy and sorrows of his people, without sacrificing the constancy of moral perfection and purpose that are basic to the divine character.[20] No one can grasp the doctrines of incarnation and crucifixion without being deeply struck by the deep intimacy of God's involvement with the world in Christ's life and death, even death on a cross (Phil. 2: 8).[21]

On the one side, the effects of the world on God will come as no surprise to a panentheist theology. A God who exists apart from his creation might remain unaffected by it. But as long as God holds the world within himself in some sense, he cannot view it with the dispassionate objectivity that the theological tradition sometimes asserted of God. Much of the tradition feared that a God who was, even in part, constituted by his reaction to the contingent decisions of his creatures would become dependent on them and thus lose some of his sovereignty. Perhaps given the philosophical tools available at the time it was unthinkable that God could truly respond to contingent beings without being essentially altered by them. But certainly no love or caring that *we humans* are able to understand – be it the love of parent, friend or lover – could exist without genuine responsiveness.

On the other side, panentheism rejects any *reduction* of God to his responses to creatures. God was before the world and will exist after it. He may possess experiences that he would not have possessed apart from its existence, such as the experience of your recent petitionary prayer; yet this fact does not make him ontologically dependent on your or my existence. In short, panentheism, like other theological approaches, maintains that the world's existence remains contingent (and God's existence necessary). At the same time, it places a particular emphasis on the fact that God genuinely responds to and is affected by what his creatures do. These two characteristics are shared by most of classical theology. In comparison to other theologies, however, panentheism has the special advantage of building this complex answer – God's responding to the world without the reduction of God *to* his response – right into its initial specification of the God/world relationship.

Difficult systematic work still needs to be done in this field of theology. How far can one go in allowing for the openness of history and the world's impact on God without sacrificing God's Lordship over history – and, in the extreme case, even the worshipability of God? Specifying parameters and excluding some options helps, I think, to 'triangulate' towards an adequate position, though a full system also demands more. On the one hand, a divine determination of history is to be rejected as incapable of doing justice to the ruptures and fissures in the terrain of human history. Full determinism – whether in its Augustinian, Anselmian or Calvinistic guise – tends to turn the stage of human history into a farce, not to mention making God responsible for evil and suffering throughout natural evolution and in the world today. Genuine openness, by contrast, allows for a genuine response. But what of the standard critique: by holding that the future is genuinely open, even if as the result of a divine self-limiting, has one not fallen into a 'limited theism', making God no longer God? Assuming that the arguments in this section are sound, we can simply bite the bullet here: if God has not determined all of human history (say, in Calvin's sense), then he simply *cannot* have foreordained every occurrence; hence he cannot be said to control every outcome.[22] Consequently, real human freedom – thus, given the fall, human sin – helps to shape human history, which thus contains events that God has not determined. There are serious conceptual reasons to think that even God cannot foreknow the outcome of genuinely free events.[23]

On the other hand, panentheism refuses to *equate* God with the becoming of the world. God existed before the (free) creation of the world and, if God is God, he will also exist after the collapse or heat death of our universe. If Christian belief is true, the final outcome of that process will involve a reconciliation of the world with God and a fulfilment of the divine purposes; the 'lure of God' will have been efficacious and not impotent. To hold Christian faith is to believe that there will finally be an eschaton and a Sabbath rest (Heb. 4: 9). But this belief is compatible with a genuine openness in the process of history leading to that rest – *and even to the possibility in principle that the result will be other than the one I believe in and hope for.* For faith the end is secure. But if human freedom is to be more than a chimera, and if God is not to be made responsible for all the ruptures along the way, then the end that I hope and long for cannot be one that already stands at the end of history like a completed structure (Pilgrim's heavenly city) – or, worse, like a corral into which all of humanity, believer and unbeliever alike, are being herded. I believe that God will triumph over evil, but I need not believe that this triumph will come about regardless of the decisions of the agents involved in that process.

Before we turn in later chapters to the theological consequences of rethinking the God/world relationship in this way, we should pause to outline the major assumptions and arguments for panentheism.

ARGUMENTS FOR AND IMPLICATIONS OF PANENTHEISM

How else can this distinction between classical Western theism and panentheism be made? What are the main reasons that point us in the direction of panentheism? The following six points present the case for panentheism, though each is actually only a shorthand sketch for what would have to be a much fuller argument.

1. The inadequacy of atheism

The theological case begins with arguments for the inadequacy of purely physicalistic or materialist accounts of the universe. Here the most important datum is perhaps the reader him- or herself. We seem to ourselves to be conscious beings, to have mental states we call feelings and thoughts which are not reducible to their physical sources, for example to neurophysiological states. The arguments against reductive materialism are complex ones; we shall return to them in Chapter 8. Without a case for the inadequacy of the purely materialist answer, the debate among theological options will of course seem arbitrary. Note that even if one assumes that an adequate metaphysics must include a place for something trans-physical, the case for the *super*natural still needs to be made, since human persons could experience mental states without anything non-natural occurring. This task will occupy us in later chapters.

Once one is convinced of the existence of a spiritual power, the crucial questions concern its relation to the world and whether it is to be understood as personal or impersonal. Although some question whether any reasons can help us to decide the latter question,[24] I do think a reasonable case can be made for the personalist answer. As Edward Pols has noted, 'The most fundamental and concrete sense of power accessible to our intelligence is power in the sense of agency.'[25] If we are trying to conceive of the divine as being (or possessing) the highest sort of power of which we are aware, we will be driven to understand the divine as agent rather than as 'merely' impersonal power. Kirkpatrick, citing Pols, argues that we will also need to make the next step: 'power in the sense of agency necessarily means "the power of an agent regarded as an entity". Therefore, if ultimacy has to do with power, then only *a* being can have the requisite ultimacy because only an agent-being can exercise power.'[26] If we conceive of God as *an agent*, we conceive of the divine as *a* being, rather than as, say, 'the Ground of Being' (Tillich). I shall speak in this way in what follows, although there are serious (and in my mind not fully resolved) metaphysical difficulties raised by speaking of the highest principle as *a* being.[27]

2. The inconsistency of classical Western theism

We earlier examined the historical forces that led the Israelites from polytheism to classical theism. What we found to be the historical development in the biblical texts also makes sense conceptually: Yahweh's relation to the world will be very different if he is thought of as one among many gods (polytheism), or as so much more powerful than the others that it is *as if* no other gods existed at all (henotheism), or as the One and Only from which all else stems (radical monotheism). I made the case that the biblical texts launch one on a trajectory of belief and thought that begs to be extended beyond the classical Western account of God's separation from the world. A theology that places God 'outside of' his creation has not yet fully thought through what is entailed in the move from many gods to the *ex nihilo* creation of heaven and earth by an absolutely infinite God. Now it would be a mistake to claim that *only* panentheism adequately fits the biblical data, for the biblical texts underdetermine the choice between systematic theological models. Nonetheless, I think it possible to show that panentheism captures the central biblical teachings about the God/world relationship, and at least a serious case can be made that it does a better job with this task than many models do, including some versions of classical theism.

3. The biblical resistance to dualism

The Israelite understanding of the human individual is radically monistic or holistic: the human being is a single entity that, although including a number of aspects, is not fundamentally divided up into separate parts.

This holistic emphasis emerges already in the opening descriptions of humanity's creation. The biblical creation texts know nothing of a pre-existent soul that is combined with – much less one that is imprisoned within! – the body. Man and Woman become individually *nephesh hayyah* or 'living soul', and no prior living soul is 'attached' to the body. Westermann adds, 'Man, created in his state as a living being, is created as a complete unity. Any understanding of man as consisting of body and soul, or of body, soul, and spirit, is excluded from the start.'[28] The notion of a soul that survives the body – and consequently any clear notion of an afterlife – does not occur until the post-Exilic period. When the belief does arise that the person survives death, there is no indication in the Hebrew texts that this post-mortem existence would be in anything other than an embodied state. The Greek dualism of mind or spirit and body, and the subsequent valuation placed upon mind over body, is thus utterly foreign to the Hebrew texts. The same is true of the New Testament: the goal for which the authors long is not a disembodied state of freedom from the body, but an integrated existence of both soul and body in the direct presence of God. The New Testament speaks not of escaping *soma* (body) but overcoming *sarx* (the flesh).[29]

The temptations toward dualism in the history of Christian theology have come primarily from the Greek notion of the priority and purity of the mind or spirit (*pneuma*) and the concurrent notion of the evilness – or at the very least, the inferiority – of the body. The Greek valuation (and devaluation) played a major role in the Patristic inference from God as pure spirit to the necessity that he be fully separate from the physical realm. The world of matter was both morally and ontologically suspect. Perhaps it possesses a lower degree of reality (Plato) or is the emanation furthest or most distant from the One (Plotinus); or perhaps it has 'privation' as its essence, as in Augustine's doctrine of evil as *privatio*; or perhaps the physical world is a complete illusion and has no reality whatsoever. At any rate, under these assumptions God must have as little to do with it as possible. And certainly he must not be identified with it. Hence the world must be external to God.

But God need not be defined as spirit *in opposition* to the world (ontological dualism) as long as matter is not seen as evil or inferior. If we take seriously the 'and it was very good' of Genesis, then it is not an evil thing for God to be closely related to the world, to 'walk in the garden in the cool of the day' (Gen. 3: 8). Indeed, if humans are really made in the image of God, imbreathed with God-given Spirit, then it would be more natural to conceive of God as Spirit working in and through at least some parts of the material world. For Christian theology, the world can never be divinised, *made into God*, but it *can* be included within the overarching span of God's universal presence and being. These are, of course, conclusions shared by many systematic theologians. Panentheism's difference is to stress the desirability of God's actually including the world, once the final vestiges of Greek or Gnostic dualism are abandoned.

4. Divine infinity

We have found that both biblical and theological lines of argument point toward the infinite/finite contrast as a crucial conceptual means for drawing the distinction between God and his creation. Yet it turns out to be impossible to conceive of God as fully infinite if he is limited by something outside of himself. The infinite may without contradiction include within itself things that are by nature finite, but it may not stand *outside of* the finite. For if something finite exists, and if the infinite is 'excluded' by the finite, then it is not truly infinite or without limit. To put it differently, there is simply no place for finite things to 'be' outside of that which is *absolutely unlimited*. Hence an infinite God must encompass the finite world that he has created, making it in some sense 'within' himself. This is the conclusion that we call panentheism.

One criticism ought to be immediately addressed: that the Bible does not explicitly speak of God as infinite. Like the doctrine of the Trinity, the doctrine of the infinity of God is an inference drawn from the biblical texts, an inference that theologians have found far more acceptable than the alternative. The Bible does emphasise the awesome power of God (Gen. 17: 1; Ps. 135: 5, 136: 6; John 1: 3), God's eternity (Ps. 102: 26f., 90: 2), God's omnipresence (Deut. 4: 39; Ps. 139: 7–12), his otherness (Isa. 46: 9), and his unknowability (Rom. 11: 33; Eph. 3: 8). Some passages come very close to a doctrine of divine infinity (Job 42; 1 Kings 8: 27, Ps. 147: 5). Augustine therefore concluded:

> It is evident that the orderly disposition of the universe comes about through a mind, and that it can appropriately be called infinite – not in spatial relations, but in power which cannot be understood by human thought . . . That which is incorporeal . . . can be called both complete and infinite: complete because of its wholeness, infinite because it is not confined by spatial boundaries.[30]

St John Damascene argues similarly:

> [God] is not to be found among beings – not that he is not but, rather, because he is above all beings and even above being itself. For if knowledge has beings as objects, then what transcends

knowledge also transcends essence and, conversely, what is beyond essence also is beyond knowledge. Therefore, the Divinity is both infinite and incomprehensible, and this alone is comprehensible about him – his very infinity and incomprehensibility.[31]

Gregory of Nyssa argues that 'if [anyone] acknowledges that the supreme Being is simple and self-consistent, then let him also grant that it combines and associates simplicity and infinity.'[32] Finally, it was Thomas Aquinas who for the first time achieved a conceptual synthesis of the infinity and perfection of God in a manner which was both philosophically adequate and theologically useful.[33] I have argued in *Das Gottesproblem* that this fusion of the two concepts, expressed in its most pithy form in Descartes' description of God as *infinita perfectio*, has not fared well in the modern period. Still, whatever problems the Scholastic metaphysics of perfection has faced, the notion of infinity – and by that I mean the *qualitative* distinction between finite or limited things and that which is absolutely infinite and hence the ground of all that exists – continues to provide a crucial resource for the doctrine of God.[34]

5. The problem of divine causality

When the world is understood as ontologically 'outside' God, then any actions that God takes within the world must represent interventions 'from outside' into the world's order. This model of God as a sort of foreign agent intervening in an independently existing order raises numerous problems. If the creation (and its Creator) were perfect from the start, why would it have to be 'fixed' in this manner? Are regularities within the world to be understood as representing a causality independent of God, one which functions all on its own? Would it not be far better theologically to view even inner-worldly causality as (in at least some sense) a manifestation of divine agency? How could these divine interventions be known at all by humans if they come from completely outside the order that we inhabit? These are not merely apologetic questions, such as the question of how one can demonstrate to the non-believer that God indeed acts within the world. Much more, they represent the urgent task of how *even believers*, speaking from the standpoint of faith, can find any way to make sense of the idea of divine 'interventions' from 'outside' into the world.

The issue of divine causality explains why panentheism and the theory of divine action are so closely linked in the present monograph. We will find in the ensuing pages that the action of God can be much more coherently conceived if the world bears a relationship to God analogous to the body's relationship to the mind or soul. Making sense of the analogy will force us to think in some detail about the relationship of the mental to the physical in humans, especially given developments in neurophysiology and gene research; and it will also require us to think carefully about God's relationship to the world. In what follows I shall call it the *panentheistic analogy*.

If the analogy turns out to be theologically fruitful, one can already begin to anticipate what it would suggest about divine action. As an opening hypothesis, it appears to suggest that there is no *qualitative* or ontological difference between the regularity of natural law and the intentionality of special divine actions. Put differently, it would seem to deny that only the latter should count as divine actions and not the former. Instead, natural laws, when viewed theologically, will count as descriptions of the predictable regularity of patterns of divine action. For example, the laws of motion, thermodynamics and gravity do not represent matters of metaphysical necessity; it would not be logically contradictory for them to be otherwise than they are, any more than it would be contradictory for fundamental constants such as Planck's constant to bear different values. The fact that our universe exhibits the physical regularities it does could be taken as a surd, a brute fact needing no further explanation (atheism); or it could be attributed to an original act of God, by which he 'set the clock in motion' and then let it run on its own (deism). Classical Western theism has held that the continuing 'concurrence' of God is required to keep things ticking along. Panentheism stands closest to classical Western theism in this regard, yet it draws an even closer link between physics and theology: since God is present in each physical interaction and at each point in space, each interaction is a part of his being in the broadest sense, for it is 'in him [that] we live and move and have our being' (Acts 17:28).

Natural regularities within God's universe, then, are roughly analogous to autonomic responses within an individual's body – the things that one's body does without conscious interference or guidance. In one sense, such behaviours are still one's own 'actions', even though they occur through the body operating in a regular or autonomic manner and one thus performs them unconsciously. For instance, one can become conscious of particular actions that one's body normally carries out automatically (by concentrating, for example, on one's breathing in and out – a process you were presumably not thinking about before reading this sentence and yet, I hope, one that you were still performing!).

101

The breathing example does suggest a difference, however: theologically, we must conclude that God, being omniscient, would *always* be aware of his autonomic or habitual actions within the universe.

I will argue later that this theological interpretation of natural regularities as (autonomic) divine actions does not rule out *conscious* actions that God might undertake. In fact the analogy with human agency actually creates the expectation that God would also consciously pursue certain ends, just as human agents also engage in actions that they perform consciously and with particular intentions in mind. How we are to understand such 'focal' divine actions and how belief in them could be compatible with the institution of science – these are questions that will demand careful attention. For now, the point is only to argue that divine action, which is obviously not an issue for atheism or deism, is better addressed within the context of panentheism than classical Western theism.

6. A closer relationship with God

Finally, panentheism conceives of an ontologically closer relationship between God and humanity than has traditionally been asserted. Theologians have of course always insisted upon the role of God as Creator, and the tradition has emphasised the necessity of a continual *sustaining* of the created order by God (the doctrine of *conservatio*). When the tradition spoke of the word of God or a sense of God within the individual human (the *sensus divinitatis*), this locution could be (and was) interpreted as the work of God in one of three different ways: as a product of the original image of God 'built into' humanity at creation and thus 'reflected in' our being in the present; as a sign of the presence of God as Conserver or Sustainer within the human being; or as a manifestation of direct divine agency. Panentheism, one might say, adds a fourth mode of the *sensus divinitatis*, an ontological one: we are aware of God because we are within God. God has created out of nothing other than himself all that exists in the finite world; we are composed, metaphorically speaking at least, out of God. More carefully, Schleiermacher wrote in the *Speeches*, 'Everything finite exists only through the specification of its boundaries, which must be simultaneously cut out (*herausgeschnitten*) of the Infinite. Only in this fashion can [each thing] be infinite within these boundaries and have a form of its own.'[35]

This way of conceiving the God/world relationship makes the relationship of Creator and created *as close as it can possibly be without dissolving the difference-in-nature between the infinite God and the finite created world*, that is without falling into pantheism. There is widespread biblical evidence of an extreme closeness in the relationship between God and the individual, to which the present view, I believe, does full justice:

> O Lord, Thou hast searched me and known me.
> Thou dost know when I sit down and when I rise up;
> Thou dost understand my thought from afar.
> Thou dost scrutinise my path and my lying down,
> And art intimately acquainted with all my ways.
> Even before there is a word on my tongue,
> Behold, O Yahweh, Thou dost know it all.
> Thou hast enclosed me behind and before,
> And laid Thy hand upon me.
> Such knowledge is too wonderful for me;
> It is too high, I cannot attain to it.
>
> (Ps. 139: 1–6)

To the Hebrew writer, God is always present within the individual human, knowing every thought and every desire and knowing it automatically and immediately – not as one listening in from the outside, but as one who is bound up with the very nature of the individual person. In the New Testament, and especially in Paul's writings, the oft-repeated phrase *en Christo* (in Christ) bears a mystical sense very similar to the phrase Luke attributes to Paul, 'in whom we live and move and have our being' (Acts 17: 28).[36] The words *en Christo* express in christological terms something of the intimate relationship between God and creation, a relationship not characterised by otherness but by the closest proximity in which two entities can exist without being identified. 'And the two shall become one' (Gen. 2: 24, Eph. 5: 31) – this central statement of the theology of marriage has a broader application as well, as Paul realised: 'This mystery is great; but I am speaking with reference to Christ and the church' (Eph. 5: 32).

Christian thought, for better or for worse, has been compelled to struggle with relations of ontological closeness that are not identities. The intimate relation of Father, Son and Spirit in classical trinitarian thought requires a distinction of persons within one identity of being, a tightrope-walking exercise that, as theologians well know, threatens all too easily to become a *salto mortales* (a death fall) into tritheism or doceticism. Similar challenges have arisen in thinking about the relation of the second person of the Trinity to the human individual Jesus (a union that the tradition takes to be complete during Jesus' life but not for all eternity) and the relation of Christ and church (a spiritual union

103

that is nonetheless less intimate than that between Jesus and the second person). A less close but still vital union is that between Christ and the believer. Torrance writes that the believer's union with Christ is onto-logical, since in the incarnation Jesus took humanity (i.e. all human nature) into himself. Yet the doctrines of sin and fall imply that this last example is less than a full union – at least until such a time as God should bring about a state of reconciliation and moral perfection currently unimaginable to humans. The tradition is thus not without tools for conceptualising the relation between Creator and created in subtle ways. Panentheism, with its insistence on principles of distinction (finite/infinite, contingent/necessary, imperfect/perfect) within an overarching oneness of being, provides an extraordinarily rich means for thinking about this relation in all its complexity.

Nonetheless, this remains difficult territory to traverse, and many pitfalls threaten the theologian who attempts to do justice to the requirements of an adequate doctrine of creation, of sameness with and difference from God. Many are the medieval theologians – I think in particular of John of Eriugena, Eckhardt, Nicholas of Cusa and Giordanno Bruno – whose attempt to speak of the closeness of the God/world relationship pushed at the limits of orthodoxy... and sometimes beyond. Nonetheless, the intimate relation described by panentheism does justice, I believe, to the biblical record as well as to the demands of conceptual consistency in a way that is not matched by most versions of classical theism. Most of all, it expresses, however partially, the individual and communal religious sense of the intimate relatedness of God and world, of God and individual, in a way more profound than traditional monotheism on the one hand and pantheism on the other.

Similarity-in-Difference

Each of the six arguments just summarised evidences a common struc-ture, clearest perhaps in the final point. Humanity by its nature – what humans are and have – is deeply, fundamentally related to the nature of God and remains a reflection of the divine nature. To possess the *imago dei* is to recognise the qualities of the divine as well as to see them as desirable. It is to long for qualities we possess only in part and to seek to be in a way in which we are not yet: 'For now we see in a mirror dimly, but then face to face; now I know in part, but then I shall know fully just as I also have been fully known' (1 Cor. 13: 12).[37] Christian theology, in contrast to much New Age thought, emphasises the difference of God from humanity[38] – as finite to infinite, contingent to necessary, reflection to source, 'in process' to actual perfection. But in their rush to preserve the God/world differences against the perceived threat of pantheism, the threat of 'making creation God', theologians have been perhaps too cautious to do justice to the fundamental ontological similarities, to the closeness of God and world. If the barriers between God and world are erected too firmly, God is marginalised from his own creation. Closer to the heart of the doctrine of God is a similarity-in-difference.

To recognise this similarity-in-difference within the notion of the *imago dei* is to see why theologians are turning to the conceptual resources of panentheism in order to understand God's relationship to the world. Humans are not 'totally other' from God but share in his being – just as they do not match either his understanding or his moral perfection but find it in their nature still to strive for it. The same similarity-in-difference is fundamental to the God/world relationship in general, and to the God/humanity relation in particular. This fact provides yet another reason why we must not approach the creation story, and in particular the fall, in the first place as a particular moment in history but rather as the archetypical expression of this relationship: 'The account of the origins', writes Westermann, 'shows in great depth and with great clarity that it belongs to man's very state as a creature that he is defective. And this defectiveness does not show itself in one single act in history, but in a variety of ways.'[39]

A panentheistic understanding of the *imago dei* and of the God/world relation has a variety of theological implications. For example, although the nature of humanity is clearly 'fallen', and thus the doctrine of sin – and consequently the doctrines of salvation and redemption – remain important, a panentheistic Christian theology stands opposed to some Reformed views of human beings as totally depraved, in the sense that the image of God is held to be completely erased from humans and the world. Here I must side with James Carpenter's brilliant treatment of the place of Augustine and Irenaeus within the theological tradition.[40] If one agrees that the biblical documents do not support a strong opposition between nature and grace (see Chapter 2 above), then it is important to understand how a major segment of the theological tradition came to set them up in strict opposition to one another.

The Scholastic principle, *gratia non tollit naturam sed perficit* (grace does not destroy nature but perfects it), sounds innocuous enough, for clearly the redemptive action of God would have to lift nature to a state more consistent with its Creator. Unfortunately, though, the principle was frequently used to support a tacit, and sometimes explicit, effort at setting nature and grace against each other as two opposed principles or realities. Thus Augustine proceeded from the principle that 'if right-eousness comes of nature, Christ died in vain' to the conclusion that 'Men go on to search out the hidden powers of nature . . . which to know profits nothing'.[41] Augustine's separation, which has had such an immea-surable impact in the history of Western theology, did not find a similar echo with the Greek theologians, who insisted upon the pervasive place of grace within nature. Irenaeus, for example, insisted that, although Adam lost the *similitudo dei* (the likeness of God or the freedom to be like God) at the fall, he did not lose the *imago dei*. Throughout history, writes Irenaeus, there have been people 'who from the beginning, according to their capacity, in their generation have both feared and loved God, and practiced justice and piety toward their neighbors'.[42] Similarly, Gregory of Nyssa notes that 'the river of grace flows every-where' and fulfils all created things.[43]

This evidence leads Philip Sherrard to conclude that the Western separation between nature and grace was responsible for 'the loss of the sense of the divine in nature . . ., a loss of the sense that the universe has a sacred quality. . ., that creation actually participates in the divine, and is an actual mode of existence or embodiment of the living, ever-present God'.[44] Carpenter likewise concludes that grace must be taken theologically 'as that which constitutes nature as well as that which restores or perfects it'.[45] Would it not be more appropriate, following Carpenter, to make theological use of the phrase, 'the grace of nature', in order to reinstate this implication of the doctrine of grace which has been so underemphasised in the Western tradition? When we speak of the 'functional unity' of nature and grace, and thus of God's creative and redemptive activity, we recover again the insight that 'all is of God and God is in all'.[46] Of course, this phrase is not to be used in the sense of pantheism, but rather in the sense of Augustine's *De vera religione*: God is the one 'of whom are all things, by whom are all things, in whom are all things'. This is the view which theologians are defending in our time under the rubric of panentheism.

RETHINKING THE SO-CALLED THEISTIC ARGUMENTS

It is interesting to note what happens as one begins to rethink the major components of the God/world relation in light of the move from classical Western theism to panentheism. In this regard it is particularly fascinating to note how the major arguments of natural theology, the so-called 'proofs of the existence of God', are transformed by the context of postmodernism and panentheism.

Three major arguments have been used since *circa* 1100 by natural theologians. Each one was originally taken to *prove* the existence of God based on one or more premises that were believed to be self-evident, with the help of a series of deductive inferences that were held to be beyond question. Given the starting premises, the existence of God was supposed to follow with deductive necessity. Such claims to deductive validity have been criticised repeatedly and soundly over the years, especially since Kant challenged the competence of reason to make any progress whatsoever in the field of metaphysics. In a post-foundationalist context,[47] it is no longer necessary to claim that natural theology can provide the platform on which the doing of systematic theology will rest.[48]

Too often, however, the critique of natural theology has led first to abandoning the major theistic arguments *in toto* and subsequently to disregarding them completely – or, in a conservative (over-)reaction against the criticisms, to a dogged preoccupation with the classic arguments and their defence. I would like to suggest instead a *theological* re-appropriation of the major arguments, taken now no longer as proofs for the existence of God but as part of the theological explication of Christian theism. That is, I would like to show how the arguments can still be 'mined' for theological insights even after claims for their probative efficacy in 'objective apologetics' have been dropped.

The Ontological Argument

The queen of the natural theological arguments has been the *ontological argument*. This argument has its classical formulation in St Anselm's *Proslogium*:

> 'the fool hath said in his heart, There is no God' [Ps. 14: 1] . . . Hence, even the fool is convinced that something exists in the understanding, at least, [namely, at least the *idea* of God] than which nothing greater can be conceived [Anselm's definition of 'God']. For, when he hears of this, he understands it. And whatever is understood, exists in the understanding. And assuredly that, than which nothing greater can be conceived, cannot exist in the understanding alone. For, suppose it exists in the understanding alone: then it can be conceived to exist in reality; which is greater.
>
> Therefore, if that, than which nothing greater can be conceived, exists in the understanding alone, the very being, than which nothing greater can be conceived, is one, than which a greater can be conceived. But obviously this is impossible. Hence, there is no

doubt that there exists a being, than which nothing greater can be conceived, and it exists both in the understanding and in reality.[49]

Anselm's first move is to claim that we do in fact have the idea of God, however limited this idea might be. He then defines God as 'the being greater than which none greater can be conceived'. This leads him to reflect on what qualities or characteristics such a being would have to have. It should be, among other qualities, all-knowing, all-powerful, all-present and all-good – or, using the 'omni's', omniscient, omnipotent, omnipresent and omnibenevolent. Were this being to lack any of these qualities, we could imagine another being who possessed them. In that case, the being we were originally imagining would be a *more limited* being, and *not* the being greater than which no greater being could be conceived. Now, what about existence? Would not a being who possessed existence be even greater (more perfect) than a being who lacked existence? Since this seems clearly true, our idea of the most perfect being must possess existence as well – which is just to say that this being must exist! Hence, Anselm concludes, from the mere *idea* of a most perfect being the actual existence of that being must follow. But we in fact have such an idea, Anselm argues, however limited our idea may be. Therefore, God must exist. (In the eighteenth century Leibniz was able to reduce this entire argument to a single-line argument: a necessary being necessarily exists. That is, from the mere idea of a necessary being – a being who exists by necessity, or a being whose very nature it is to exist – it follows that that being must *actually* exist.)

Now this proof, however intriguing, has been widely criticised. Many find Kant's classic criticism in *The Critique of Pure Reason* persuasive: 'Existence is not a predicate'. It is a category mistake, according to Kant's charge, to place existence alongside omnipotence and omniscience as one of the 'attributes' of the idea of a perfect being. Indeed, it is illicit to move from the idea of *anything* to its existence, because no matter how many qualities, of whatever sort, you add to the 'idea' side of the equation, it can never propel you across the great divide to 'hence, it exists'.

Kant notwithstanding, the ontological proof is by no means dead philosophically. A number of Christian thinkers have used the new resources of the 'logic of possible worlds' (modal logic) to renew the case for the ontological argument,[50] and it plays a major role in the 'Anselmianism' of many of the members of the Society of Christian Philosophers and their publications.[51] It is not our task to resolve this debate here, but rather to ask what can be learned about the God/world relation from the ontological argument as classically formulated. The argument turns on our sense of a being (or a level of being) whose nature is radically different from our own. God, if he exists at all, exists of necessity – by the necessity of his own nature since, as St Thomas wrote, God is *ens necessarium* (necessary being). It would be a contradiction to say that one has the true idea of God and yet imagines God as not existing – whether or not that contradiction has any probative force.

Theologically, the 'ontological' line of reflection gives expression to the most fundamental difference imaginable between God and creatures: the modal difference. Prior to any other quantitative differences in the two kinds of being (e.g. their degree of goodness, of knowledge or of power) is the modal status of the beings. There are three basic choices here: an entity or state of affairs might be impossible (e.g. a round square, a world in which two plus two equals five); it might be possible (a man or woman who could run a three minute mile, an Irishman named Nancy Slovanitch, a unicorn); or it might be necessary, that is true in all possible worlds (e.g. $7 + 5 = 12$).[52] Every other existing thing in the universe, and even presumably the universe as a whole, belongs to the second category (obviously nothing that exists could belong to the first category!). God alone, however, belongs to the third category. (Actually, several entities could exist of necessity, and many truths could be necessarily true. Still, only one entity could be both necessary and all-powerful, since two such beings would end up limiting each other's power.) Even for the panentheist, the first observation about human existence is that we are contingent – we might not exist – whereas the God who is our Source exists necessarily and is thus radically Other than us. Note that this observation, as I treat it here, is not a valid proof that God must exist, as the tradition thought, but rather a way *to explicate the logic of the notion of God*.[53] This modal belief about God is, conceptually speaking, one of the most fundamental ways of expressing his divinity as over against the being of all created things. Many of the other, more specific distinctions between God and creatures can be derived from this fundamental distinction, and all others presuppose it.

The ontological argument is also consistent with panentheism. The individual thinking subject is enough like God to possess the idea of God. It is in our nature to have the idea of a being greater than which none greater can be conceived.[54] Recall Anselm's quotation from Psalm 14, 'even the fool hath said in his heart, There is no God.' Certainly it is not in our nature to *be* this being, or even to have a perfect idea of it. The idea of God is built into us, yet far transcends us; we are similar enough to God in our nature to think God, yet not so similar to fully comprehend what it is that we are thinking. We are God-oriented, God-directed; we possess minds that incline naturally toward immortality; yet we do not possess immortality. We long for what we are not; we incline toward a being whose nature it is to be qualitatively different from ourselves; we have, to paraphrase Pascal, a God-shaped hole in our hearts and our minds which only a knowledge of God can fulfil – these are the sorts of theological insights to which the ontological proof contributes.

The Cosmological Argument

The *cosmological argument* begins, in its most simple form, from the single empirical premise, 'Some contingent things exist.' St Thomas gave the argument its classical formulation in his famous 'third way', the so-called argument from possibility and necessity:

> We find in nature things that are possible to be and not to be, since they are found to be generated, and to be corrupted, and consequently, it is possible for them to be and not to be. But it is impossible for these always to exist, for that which can not-be at some time is not. Therefore, if everything can not-be, then at one time there was nothing in existence. Now if this were true, even now there would be nothing in existence, because that which does not exist begins to exist only through something already existing. Therefore, if at one time nothing was in existence, it would have been impossible for anything to have begun to exist; and thus even now nothing would be in existence – which is absurd. Therefore, not all beings are merely possible, but there must exist something the existence of which is necessary. But every necessary thing either has its necessity caused by another, or not. Now it is impossible to go on to infinity in necessary things which have their necessity caused by another, as has been already proved in regard to efficient causes. Therefore we cannot but admit the existence of some being having of itself its own necessity, and not receiving it from another, but rather causing in others their necessity. This all men speak of as God.[55]

If something contingent exists, argues Thomas, this means that it does not possess its reason for existing within itself (as a necessary being would) but rather receives it from something else; we speak of this other thing as its 'cause'. Now what caused the contingent thing (say, you) must itself be either contingent or necessary. If the former, it too must be caused by something outside of itself. But this would throw us immediately into a regress of causes, leading back further and further into history (the parents of the parents of the parents of . . .). Should the regress be infinite – an unending series of contingent beings or states, each caused by something before it – then there is no reason for the existence of *any* of the beings; the existence of each particular being, and indeed the existence of all that is, is just a brute fact depending on the prior existence of other brute facts. The other possibility, however, is that the regress of causes is *finite*. But this would happen only if the first, originating cause had its reason for existing within itself – that is, only if it were a necessary being. For only if there is a necessary being at the beginning of the whole chain of contingent beings will their existence have a 'reason' or an explanation. But there *must* be such a reason; hence, Thomas concluded, God must exist.

The classical argument as found in St Thomas and Leibniz used this line of reasoning as an argument for God's necessity. Thus Leibniz formulated the Principle of Sufficient Reason: *for everything that exists, there must be a reason why it exists rather than does not exist.* Recent defenders have tried other modifications in the argument. For example, in a famous objection Bertrand Russell argued that there is no need for a first cause because the series of causes might be infinite, and an infinite series does not have, or need, a first cause. Defenders have then replied that, although an infinite series would not have a first cause, one would still have to give the reason why *this infinite series* exists rather than does not exist. If *that* reason or cause is merely contingent, we are launched into *yet another* infinite regress of causes, for we would then have to ask again why the cause of the infinite series exists, and why *its* cause exists, and so on *ad infinitum*.

Even given recent developments in the literature on the cosmological argument, it seems clear that it does not supply us with a knock-down proof of the existence of God. Critics have charged that the argument confuses reasons and causes, and they have insisted that there is no reason to think that the Principle of Sufficient Reason (for everything that exists, there must be a reason why it exists rather than does not exist) must be true – unless one arms herself with a belief in the existence of God in the first place![56] Nonetheless, the cosmological proof remains rich in theological insight. Among other things, it helps one to think through the implications of divine being as one that exists by the necessity of its nature alone. As we have already seen, only one God, one all-powerful necessary being, could exist; hence God must be the *sole* source of all that is (cf. radical monotheism). Also, God must be eternal, since he couldn't both be necessary and come into existence at some point. Finally, the argument expresses many of the key parameters for any doctrine of creation. As the source for all contingent things, God supplies the full or 'sufficient' explanation of all he creates; to know God is to know why all things are here. It is also to know what the nature of things is – to know them *as created*.

The argument also powerfully expresses the relation of *ontological dependence*: it is in the nature of contingent things to depend for their existence upon a necessary being, one who is the source of the entire order or series of causes that has led to their presently existing. Here, too, the theological extrapolations lie close at hand. It is in the nature of a contingent being not only to owe its existence to God *indirectly*, as mediated through the previous contingent causes that lie between it and God. It is also in its nature to owe its ongoing existence to God *directly*, in so far as the continuing existence of the entire chain of causes must be due to God's constant concurrence or grace. Moreover, the argument suggests, the original creative act of God could not have occurred by necessity, since this would transfer God's necessity onto the creation itself, which would thereby lose its contingency. Instead, it must have been a *free* decision and act on the part of God that gave rise to a universe of contingent beings in the first place. Note further that a free decision of this sort entails that God be understood as a conscious or thinking being rather than as a principle or mindless power working out the necessity of its own nature. Finally, the natural response of a contingent being is to recognise its dependence on its source. That recognition is central to what the Jewish and Christian traditions have called *worship*, which is most fundamentally the act of recognising the divinity of God and the dependence of the creature upon him.

The Teleological Argument

Also called 'the argument from design', this classical proof is much easier to follow. Its classic formulation can be found in William Paley's *Natural Theology* of 1802:

> But suppose I had found a *watch* upon the ground, and it should be inquired how the watch happened to be in that place, I should hardly think of the answer which I had before given, that for anything I knew the watch might have always been there. Yet why should not this answer serve for the watch as well as for the stone; why is it not as admissible in the second case as in the first? For this reason, and for no other, namely, that when we come to

inspect the watch, we perceive – what we could not discover in the stone – that its several parts are framed and put together for a purpose, e.g., that they are so formed and adjusted as to produce motion, and that motion so regulated as to point out the hours of the day.... [57]

Now eyes are complicated mechanisms, like watches or cameras. Watches and cameras are the products of intelligent design by conscious beings. Since eyes are analogous to them, eyes too must be the product of intelligent design. Clearly they were not designed by any human designer; hence they must have been designed by God.

More complicated versions of the argument exist in the literature, of course. Defenders have needed more sophisticated versions in response to the 'Darwinian objection', which claims that evolutionary theory can fully explain the existence of complex life forms through natural mechanisms – random genetic variation and selective retention through the struggle for survival – rather than through a Creator's design. Recent defences thus emphasise the need for *meta*-order: orderly laws that regulate the process of biological development and are themselves best explained by the existence of God. But what matters for present purposes is the fundamental intuition behind all such teleological arguments: complex and highly ordered phenomena within the universe, and especially within the biological world, must be the product of design. Hence God must exist.

This argument, unlike the previous two, is an argument by analogy. That means that it can only show, at best, that life is *probably* the result of intelligent design. The strength of the argument turns on the strength of the analogy. If eyes (or the circulation of blood through the body[58]) were overwhelmingly like watches or cameras, and if we lacked any *more* credible explanation of how there came to be eyes, then we should affirm that eyes were probably designed by a very powerful intelligence. But the argument faces troubles on both fronts: there are some significant disanalogies, and most biologists believe that we already possess a good explanation of how eyes or other organs came to evolve. Moreover, there are significant cases of *disorder* in the biological world, including the pointless death of thousands of species. Should not the cases of disorder count against the hypothesis of a cosmic designer? We owe to David Hume perhaps the most powerful critique of the teleological argument, and the most rhetorically effective:

This world, for aught [the teleologist] knows, is very faulty and imperfect, compared to a superior standard; and was only the first

113

rude essay of some infant deity who afterwards abandoned it, ashamed of his lame performance: It is the work only of some dependent, inferior deity, and is the object of derision to his superiors: It is the production of old age and dotage in some superannuated deity; and ever since his death has run on at adventures, from the first impulse and active force which it received from him. . . . And I cannot, for my part, think that so wild and unsettled a system of theology is, in any respect, preferable to none at all.[59]

The teleological argument has been perhaps the most roundly criticised of any of the proofs, and yet it remains perhaps the most enduring of them all. How can this be? *To the eyes of faith*, it is impossible not to find signs of God's purpose and design within the world. These signs cannot function as a proof, compelling belief; it is even difficult to show in an objective fashion that theism provides a *better* explanation of the biological data than does neo-Darwinian evolutionary theory.[60] Instead, the erstwhile teleological proof becomes a step in the process of *fides quaerens intellectum*, faith seeking understanding. *If* God exists, then God's existence *must* be, in one way or another, part of the explanation for the complicated structures we find in the universe. Conscious beings cannot have arisen by chance in a universe created by a providential God. The believer, looking at the scientific facts, cannot but see the hand of God in and behind them; and she would be remiss as a believer were she *not* to recognise signs of the God whom she believes to be sustaining and guiding the process of universal history.

THE GOD OF INFINITE PERFECTION AND THE GOD OF JESUS CHRIST

The previous two chapters explored the Israelite and Christian doctrines of God's relation to the world. There we found a God who was actively involved with the creation, who sought to reproduce the divine image (or something like it) within creatures, who was good by nature and who sought to develop within humankind a worshipful and obedient response to its Creator and an ethical response toward others. We found a God who continued to be redemptively involved with this creation, one who was responsive to humanity's changes in world-view and values, keeping pace with human evolution like a trainer running alongside a younger athlete. In the previous chapter we traced the transformation of Israelite monotheism through the doctrine of the incarnation of God in Jesus Christ. In this chapter we then stepped back to engage systematic concepts of God. Working in dialogue with the biblical documents and with later reflection upon them, we found ourselves driven to formulate a panentheistic notion of God as both transcending and including the world. Finally, we reread the traditional Western arguments for the existence of God in light of the post-foundationalist (or 'postmodern') context. The content that emerges when the erstwhile proofs are read theologically – i.e. as expressions of the logic of Christian belief – is again consistent with panentheism.

A first exploration of the biblical texts is thus behind us and the explicit discussion of scientific results before us. It is thus a good time to ask: what is it that the 'conceptual disciplines' – natural theology, philosophical theology and metaphysics – have contributed to the theology of the God/world relation? Though much could be said here, five summary points will have to suffice:

1. *They provide general conceptual criteria for any doctrine of God.* Any theory of God must be consistent (lest one not even know what one is asserting!), coherent and adequately comprehensive; it must also have some explanatory power, for instance in explaining human nature and the cosmos in which we find ourselves. That the philosophical disciplines provide some criteria for testing theological theories does not mean that theology is shackled to – or, God forbid, subordinated to – philosophy. At the very least, as Wolfhart Pannenberg has often argued, the conceptual disciplines provide formal standards for any adequate theory of God.[61]

2. *They provide models of God that can be filled with theological content, as well as examples of past mistakes that theologians must avoid.* To impose on systematic theology a complete and ready-made theory of God into which, like a Procrustean bed, the biblical data would have to be folded would be methodologically unacceptable. But broader conceptual models for conceiving the divine nature *can* be helpful to the theologian as she puzzles through the diversity of the biblical narratives and the theological tradition itself. Several authors have recently shown how fruitful the notion of multiple models can be for the doing of systematic theology.[62]

3. *They provide a clear picture of what occurs when certain biblical or theological motifs are thought through in a systematic and rigorous manner.* One excellent example can be found in the history of the metaphysics of perfection, alluded to above. It seemed like a natural step to move from the biblical passages about God's goodness to ascribing absolute goodness to God. In the twelfth century St Anselm then suggested that one think of God as Goodness Itself, which St Thomas also understood as one of the *transcendentalia* of God.

115

Descartes then worked to conceive God in a consistent sense as *infinita perfectio*, infinite perfection. His work launched a 150-year tradition of *ontotheology*, which culminated in the most rigorous metaphysics of perfection of the modern period, the philosophy of G. W. Leibniz. But this attempt to conceive divine perfection as the best explanation for the world as we know it gave rise to what appear to be insuperable conceptual problems. The further the modern West moved from a pre-scientific or 'Aristotelian-Thomistic' world-view, according to which all things strive for the state (and the location) of maximum perfection, the more pronounced the problems became. The metaphysics of timeless essences and of universal striving for perfection came to stand in deep conflict with the conceptual world of modern science.

Now one might conclude that the metaphysics of perfection is simply intrinsic to Christianity, arguing 'so much the worse for science' or 'so much the worse for theology', depending on one's own proclivities. But – assuming that this conflict between modern science and the metaphysics of perfection is indeed a fundamental one – one can also begin to question whether theology has hooked its cart to the wrong horse. One alternative[63] is to look to metaphysical options, such as the *Ethics* of Spinoza, which offer a doctrine of God without a metaphysics of perfection. Within the Spinozistic tradition 'God' became the name for the infinite dimension within nature or, in later Spinozists, for its infinite source. Theologians will of course wish to correct for the pantheism in Spinoza's metaphysics — but this is a correction which eighteenth-century thinkers have already carried out for us *on the basis of internal difficulties in Spinoza's thought itself*. One important correction can be found in Schleiermacher's early work, especially the *Speeches*, another in Friedrich Schelling's 1807 *Essay on Freedom*. Schleiermacher presents a Romantic theology of the infinite, Schelling a consistent theology of divine and human freedom.[64] Both represent progress over Spinoza's own view, and both face difficulties of their own. Theologians who are aware of these metaphysical developments will be able to appropriate their successes and will never be doomed to repeat their failures.

4. *They allow us to formulate the alternatives to Christian theology in a rigorous fashion and thereby to indicate areas of disagreement and of potential agreement.* Christian theology has, for example, feared pantheism since its earliest centuries. But what happens when pantheism is formulated as a careful position and explored in an open manner to discover what it really has to offer?[65] We have seen that the

attraction of pantheism is twofold: it stems from a rigorous interpretation of what it means for God to be truly infinite, and it reflects an emphasis on the mystical or experiential approach to God in contrast to the more reflective or theoretical approach.[66] Once one has come to understand the position, its weaknesses become clearer. One reservation we have already encountered stems from the observation that speaking of the highest principle as personal represents a higher category than construing it merely as an impersonal force. If this is true, the attributes that we connect with the personal (intention, will, choice, rationality and the like) will be indispensable parts of a religiously adequate metaphysic. That the word 'pantheism' continues to have the word 'theism' within it expresses the mystical sense of some sort of religious connection with the world. Still, one must be careful not to attribute personal attributes to the one all-that-is which are denied by the pantheistic position itself.

When one explores a position intimately in this fashion, one is also in a better position to understand its rational credentials and its continuing attractiveness. For example, the widespread interest in 'spirituality' today reflects the ascendence of pantheistic ways of thinking in recent years. Not since the Romantic movement in the early nineteenth century has pantheism been such a popular position. New Age religious thought is deeply pantheistic, and its close affiliation with Native American and Eastern religious traditions confirms this fact. The popularity of these forms of thought has led to a conservative counter-reaction that borders on the *ad hominem*:

> Pantheism finds fertile soil in egocentric Western culture. Its message declares what a self-centered society wants to hear. This makes pantheism particularly dangerous, for it caters to our narcissism. Should we worship ourselves?[67]

Is Christianity compatible with pantheism? Ultimately, I do not think so; but the level of the entire discussion would be raised and unnecessary dichotomies would be eliminated if its actual strengths were acknowledged. For, even if pantheism 'goes too far' for Christian theology, it stems from motivations that are not utterly foreign to the biblical texts and the theological tradition. One of those is reflected in the well-known passage from Acts, 'In him we live and move and have our being' (17: 28). Like the repeated *en Christo* in the Pauline epistles, this phrase connotes the basic theological understanding that there is being or existence in Christ alone, that outside of God there is no life, thought or hope. The dangers of 'going too far' must not be allowed to obscure the abiding theological point: our life is in God, and apart from him we have no life at all. God's ultimacy and infinity have theological priority; we are *ex nihilo*, out of nothing; we are nothing without him.

Yet we are not God. As basic to Christianity as the all-sufficiency and ultimacy of God is the 'more than' of God. In comparison with pantheism, panentheism understands humans and the world to be both *more* and *less*. Less, for we are not God but creation, not necessary but contingent, not ultimate but derived. And yet more: in pantheism all things that appear flow from God and are destined eventually to flow back into God – like the Hindu (*advaita vendanta*) metaphor of the person who, like a drop of water returned to the endless sea, loses all distinction and becomes one with the infinite All of which it is part. In the Christian view the soul is eternal; even in a final state of the 'unity of all in God', individual souls or persons remain, united with God and yet distinct from him. This is the Christian hope and belief, that this 'moment' of universal history, this slice of a few billion years when conditions are right for higher life forms to exist at all, is not a merely passing moment, but one with eternal consequences. This hope, however much at variance with the 'fry or freeze' alternative offered us by contemporary cosmology, is basic to the Christian message. It finds its expression as much in the metaphysical debate with pantheism as it does in the cosmological debate with the 'heat death' of the universe or the 'Big Crunch', the two dominant theories in contemporary physics.

Incidentally, just as pantheism remains a continuing temptation, so too does deism. Deism – the belief in the non-involvement of God in the world subsequent to creation – represents the backdrop for the entire discussion of divine action in the world today, as later chapters will show. (In fact, we will find that deism and pantheism have important similarities and often make good allies.) The problems with divine causality that have emerged in the scientific age make it essential that Christian theologians understand the logic of the deistic response as fully as they understand their own tradition.

5. *Finally, they turn our attention back to the true universality of the claims made by the Christian narratives about God.* There can be no question of providing universal arguments that (as the tradition held) 'compel the assent of all men'. But the demise of objective apologetics does not affect the breadth of the story that lies at the heart of Christian

theology. What type of reflection is appropriate once one has been directed by one's examination of the philosophical arguments to look toward *history* in hopes of discovering the concrete actions of a free, personal God? It would seem appropriate then to turn away from, or at least to relativise, a priori reflection, focusing instead on the flow or pattern of history as a whole – and especially on those events within history where the probability seems highest of detecting some signs of divine activity. Surely the major religions of the world and their histories would be an area to study carefully in carrying out this task, as an increasing number of theologians are coming to recognise. This implication is an important one and bears restating: the same motivations that lead Christian theologians to pay close attention to metaphysics should also lead them into an open and sustained study of the various religious traditions of the world.

One of the advantages of continuing the discussion with the philosophical disciplines and with work in comparative religions, then, is that it keeps the theological focus on the world-as-a-whole and on universal history as the actual *locus* of divine activity. Redemption, salvation, sanctification – none of these Christian doctrines is about the transformation of the individual alone, or even of the group of individuals who constitute the church. The Christian story is finally about the redemption and transformation of the world, about universal history, and about a hope beyond the end of the world.

In this chapter we have made the transition from the biblical texts to a more philosophical consideration of the doctrine of God. The core argument aimed to show that panentheism does justice to the movement from polytheism to monotheism, and in fact represents a *more adequate* completion of this movement than was classical Western theism. At the same time, panentheism remains faithful to the central impulses behind classical theism as they are found in the biblical documents and in the theological tradition. It is also better able – so the thesis of the present work – to address the problem of divine agency in our contemporary scientific and philosophical context.

NOTES

1. See Thomas Bulfinch, *Bulfinch's Mythology* (New York: Modern Library, 1993).
2. H. Richard Niebuhr, *Radical Monotheism and Western Culture* (New York: Harper Torchbooks, 1943, 1960), pp. 32, 37.

3. See Bill Devall and George Sessions, *Deep Ecology* (Salt Lake City, UT: G. M. Smith, 1985), especially Sessions' appendix on Spinozism, 'Western process metaphysics (Heraclitus, Whitehead, and Spinoza)'. Also see George Sessions (ed.), *Deep Ecology for the Twenty-First Century* (New York: Random House, 1995); and Alan Drengson and Yuichi Inoue (eds), *The Deep Ecology Movement: An Introductory Anthology* (Berkeley, CA: North Atlantic Books, 1995).

4. See Jean-Luc Marion, *God Without Being: Hors-Texte*, trans. Thomas A. Carlson (Chicago: University of Chicago Press, 1991).

5. Jürgen Moltmann, *God in Creation: A New Theology of Creation and the Spirit of God* (Minneapolis: Fortress Press, 1993).

6. See the excellent discussion in Wolfhart Pannenberg, *Systematic Theology*, Vol. 2, trans. Geoffrey W. Bromiley (Grand Rapids, MI: Eerdmans, 1991).

7. See Pannenberg, *Systematic Theology*, Vol. 2, Chapter 7, 'The creation of the world', especially pp. 79f.

8. Moltmann, *God in Creation*, p. 154.

9. Moltmann, *God in Creation*, p. 155, emphasis added.

10. For more detail see Philip Clayton, 'The theistic argument from infinity in modern philosophy', *International Philosophical Quarterly* 36 (1996): 5–17.

11. Claus Westermann, *Creation*, trans. John Scullion, SJ (Philadelphia: Fortress Press, 1974), p. 19.

12. See Yeats, 'The Second Coming', in M. L. Rosenthal (ed.), *Selected Poems and Two Plays of William Butler Yeats* (New York: Macmillan, 1962), p. 91.

13. Indeed, Bonhoeffer sees the first sign of this ethical break already in Genesis 3:7, in which Man and Woman feel shame as they notice the new way in which they are being stared at by their partner and seek to cover themselves: 'Man demands his portion of the body of the woman; more generally, one person demands "his" portion from the other, raising a claim to possess the other [or a portion of him] and thereby denying and destroying the createdness of the other' (Dietrich Bonhoeffer, *Creation and Fall: A Theological Interpretation of Genesis 1–3* (New York: Macmillan, 1959), p. 99).

14. 'If the narrator has deliberately used the same formulas in the two questions which God puts to the guilty parties: "Adam, where are you?" and "Where is Abel, your brother?", so that one echoes the other, he does it so as to acknowledge that what he calls sin, what we designate as man's limitations, embraces a human perversity directed both against God and against one's brother. The narrator makes these questions correspond so as to point out that man's responsibility to God and his responsibility in community cannot be separated' (Westermann, *Creation*, p. 20).

15. This was the main theme of my *Das Gottesproblem: Gott und Unendlichkeit in der neuzeitlichen Philosophie* (Paderborn: Ferdinand Schöningh Verlag, 1996); in English as *Infinite and Perfect? The Problem of God in Modern Philosophy* (forthcoming).

16. The effect of such multiple 'models' of God is worked out effectively in Sallie McFague, *Models of God: Theology for an Ecological, Nuclear Age* (Philadelphia: Fortress Press, 1987). I believe a similar argument underlies

Jürgen Moltmann's reference to the Holy Spirit as 'she' in *God in Creation*.

17. Perhaps better than any other in the twentieth century, Wolfhart Pannenberg has noted, explained and incorporated this transformation of classical thought in the direction of a metaphysics of historicity. Since 1963 his discussions with process thinkers have informed his own systematic theology – at the same time that he has maintained a distinction between process thought and his own. See especially the contributions by the process thinkers in Carl Braaten and Philip Clayton (eds), *The Theology of Wolfhart Pannenberg: Twelve American Critiques* (Minneapolis: Augsburg, 1988) and Pannenberg's responses.

18. See Alfred North Whitehead, *Process and Reality*, corrected edition, ed. David Ray Griffin and Donald W. Sherburne (New York: Free Press, 1978).

19. Perhaps most relevant here is Charles Hartshorne's *A Natural Theology for Our Time* (La Salle, IL: Open Court, 1967).

20. For a balanced treatment of this subject see Paul Fiddes, *The Creative Suffering of God* (New York: Oxford University Press, 1988).

21. See Jürgen Moltmann, *The Crucified God: The Cross of Christ as the Foundation and Criticism of Christian Theology* (New York: Harper & Row, 1974).

22. It is encouraging to find theologians working within the classical traditions who are also willing to accept consequences such as these. See, for example, Clark Pinnock et al., *The Openness of God: A Biblical Challenge to the Traditional Understanding of God* (Downers Grove, IL: InterVarsity Press, 1994).

23. For a full discussion of the arguments see John Martin Fischer, *The Metaphysics of Free Will: An Essay on Control* (Cambridge: Blackwell, 1994), and Fischer (ed.), *God, Foreknowledge, and Freedom* (Stanford: Stanford University Press, 1989).

24. See especially John Hick, *An Interpretation of Religion* (New Haven, CT: Yale University Press, 1989).

25. Edward Pols, 'Power and agency', *International Philosophical Quarterly* 11 (1971): 295f.

26. Frank G. Kirkpatrick, 'Understanding an act of God', in Owen Thomas (ed.), *God's Activity in the World: The Contemporary Problem* (Chico, CA: Scholars Press, 1983), p. 165.

27. I explore some of these difficulties in *Das Gottesproblem*.

28. Westermann, *Creation*, pp. 77–8.

29. See Robert H. Gundry, *Soma in Biblical Theology: With Emphasis on Pauline Anthropology* (New York: Cambridge University Press, 1976).

30. Epistle 118, quoted in Leo Sweeney, SJ, *Divine Infinity in Greek and Medieval Thought* (New York: Peter Lang, 1992), p. 8. Professor Sweeney's collection of essays is unrivalled as an exhaustive history of the concept of the infinity of God in Patristic and medieval theology.

31. *De fide orth.* 1. 4, quoted in Sweeney, *Divine Infinity*, pp. 8f.

32. Nyssa, *Contra Eunomium*, Book One, C. 19, PG 45, 324, quoted in Sweeney, *Divine Infinity*, p. 484.

33. Aquinas, *In I Sent.*, 43.1.1; ST, 1a, 7. 1–2.

34. See Robert J. Russell, 'The God who infinitely transcends infinity: insights

from cosmology and mathematics into the greatness of God', CTNS [*Center for Theology and the Natural Sciences*] *Bulletin* 16 (1996): 1–13. I examine the theological contribution of a metaphysics of infinity from Schleiermacher and Hegel to the present in *Das Gottesproblem*, Vol. 2, *Moderne Lösungsversuche* (Paderborn: Ferdinand Schöningh Verlag, forthcoming).

35. 'Alles Endliche besteht nur durch die Bestimmung seiner Grenzen, die aus dem Unendlichen gleichsam herausgeschnitten werden müssen. Nur so kann es innerhalb dieser Grenzen selbst unendlich sein und eigen gebildet werden.' See Schleiermacher, *Über die Religion*, p. 53 of the 1799 edition, quoted here from the Otto Braun edition, Philosophische Bibliothek, Vol. 139b (Leipzig, 1911).

36. See George Ladd, *A Theology of the New Testament*, revised edn, ed. Donald A. Hagner (Grand Rapids, MI: Eerdmans, 1993). Maximilian Zerwick, SJ (*Biblical Greek*, trans. from the 4th Latin edn by Joseph Smith, SJ (Rome: Scripta pontificii instituti biblici, 1963), p. 40) notes that the Greek preposition *en* in the phrase *en Christo*, like the genitive, 'may . . . express even the most exalted mystical union between Christ and those who are "Christ's". . .'

37. We find the same structure in the tree of life and the tree of the knowledge of good and evil in the Garden myth: 'The insertion of the motif of the tree of life is to be explained as follows: wisdom (in the sense of an all-embracing knowledge) and eternal life belong to the divinity. But man is created with a longing for knowledge and with a longing for life. Striving after knowledge and after a life which is indestructible is a question of belonging to the divine, of being like God, as is presented in the temptation scene' (Westermann, *Creation*, p. 106).

38. See Ted Peters, *The Cosmic Self: A Penetrating Look at Today's New Age Movement* (San Francisco: Harper, 1991); and David K. Clark and Norman L. Geisler, *Apologetics in the New Age: A Christian Critique of Pantheism* (Grand Rapids, MI: Baker Book House, 1990).

39. Westermann, *Creation*, p. 121.

40. See James A. Carpenter, *Nature and Grace: Toward an Integral Perspective* (New York: Crossroad, 1988).

41. Augustine, *On Grace and Free Will*, 25, and *Confessions*, X.xxxv.55, quoted in Carpenter, *Nature and Grace*, pp. 7, 5.

42. Irenaeus, *Against Heresies*, ID.xxii.2, quoted in Carpenter, *Nature and Grace*, p. 23.

43. Gregory of Nyssa, *Oration on Baptism*, 46, quoted in Carpenter, *Nature and Grace*, p. 30.

44. Philip Sherrard, 'The desanctification of nature', in Derek Baker (ed.), *Sanctity and Secularity* (Oxford: Basil Blackwell, 1973), pp. 1–2.

45. Carpenter, *Nature and Grace*, p. 164.

46. Carpenter, *Nature and Grace*, pp. 184–5.

47. See Wentzel van Huyssteen, *Essays on Postfoundationalist Theology* (Grand Rapids, MI: Eerdmans, 1997) and the discussion in Chapter 1 above.

48. I have argued this in detail in *Infinite and Perfect? The Problem of God*. It is

interesting to note that very few systematic theologians would accept a foundational role for natural theology – even among those theologians who are often accused of being 'rationalistic' in their approach. Thus, for example, Wolfhart Pannenberg eschews natural theology in this sense, at the same time that he holds that the continuing dialogue between theology and philosophy is crucial.

49. See Anselm, *Proslogium*, chapters 2–4, quoted from Baruch A. Brody (ed.), *Readings in the Philosophy of Religion: An Analytic Approach*, 2nd edn (Englewood Cliffs, NJ: Prentice Hall, 1992), pp. 98ff.

50. See especially Alvin Plantinga, *The Nature of Necessity* (Oxford: Clarendon Press, 1974), particularly Chapter 10.

51. See Thomas Morris, *Anselmian Explorations* (Notre Dame, IN: University of Notre Dame Press, 1983).

52. Modal differences are thus differences in logical status, as in the differences between something being necessarily true, possibly true or impossible (respectively, '5 + 7 = 12'; 'It's snowing'; and 'This square is round').

53. For more detail, see Wolfhart Pannenberg, *Metaphysics and the Idea of God*, trans. Philip Clayton (Grand Rapids, MI, Eerdmans, 1990).

54. See the 'Cartesian mediation' that opens Chapter 3 of Clayton, *Das Gottesproblem*.

55. Thomas Aquinas, *Summa Theologica*, Part I, Question 2, art. 3, translated by the Dominican Brothers of English Provence (New York: Benzinger, 1947).

56. See, for example, Patterson Brown, 'Infinite causal regression', and William Rowe, 'Two criticisms of the cosmological argument', both conveniently reprinted in Baruch Brody (ed.), *Readings in the Philosophy of Religion: An Analytic Approach*, 2nd edn (Englewoods Cliffs, NJ: Prentice Hall, 1992), pp. 123–37 and 142–57.

57. William Paley, *Natural Theology* (1802), reprinted in Baruch Brody (ed.), *Readings*, p. 158.

58. It is holistic biological systems such as that of blood circulation that have led some biologists to scepticism about Darwinian evolution without divine guidance and intervention. See in particular Michael J. Behe's recent book, *Darwin's Black Box: The Biochemical Challenge to Evolution* (New York: Free Press, 1996).

59. David Hume, *Dialogues Concerning Natural Religion*, Part VI, ed. Richard Popkin (Indianapolis: Hackett, 1980), pp. 37f.

60. Although I also attempt something like this in 'Inference to the best explanation', *Zygon* 32 (1997), 177–91.

61. See, among numerous works, Pannenberg, *Metaphysics and the Idea of God*.

62. See Sallie McFague, *Models of God: Theology for an Ecological, Nuclear Age* (Philadelphia: Westminster, 1987); Philip Clayton, *Infinite and Perfect? The Problem of God in Modern Philosophy* (forthcoming); Ian Barbour, *Myths, Models and Paradigms: A Comparative Study in Science and Religion* (New York: Harper & Row, 1974).

63. See Clayton, *Das Gottesproblem*, Chapters 7–8.

64. See Clayton, *Das Gottesproblem*, Chapter 9.

65. For an excellent example, see Michael P. Levine, *Pantheism: A Non-theistic Concept of Deity* (New York: Routledge, 1994).

66. This is not to say that theism does not leave room for, and even demand, a mystical dimension as well. Conversely, some have claimed that pantheism is the ultimately reasonable stance. Thus John Hunt writes, '[Pantheism] is the theology of reason, of reason it may be in its impotence, but still of such reason as man is gifted with in this present life. It is the goal of Rationalism . . . because it is the goal of thought' (*Pantheism and Christianity* (Port Washington, NY: Kennikat Press, 1884, 1970), p. 390).

67. David K. Clark and Norman L. Geisler, *Apologetics in the New Age: A Christian Critique of Pantheism* (Grand Rapids, MI: Baker Book House, 1990), p. 223. With the exception of a few such comments, it must be said that Clark and Geisler present a well-reasoned and carefully argued presentation and critique of pantheism, one characterised by and large by argumentation rather than prejudgement.

PART II

———ᴧᴧᴧᴧᴧ₨⊚₨ᴧᴧᴧᴧᴧ———

The New Scientific Context

5

CREATION AND COSMOLOGY: WHAT THEOLOGIANS CAN AND CANNOT LEARN FROM SCIENTIFIC COSMOLOGY

————◦◦◦◦◦◦◦◦◦————

What does it look like to do theology with one eye to the results of actual scientific inquiry? Can a Christian theologian incorporate the results of science without sacrificing what she views as essential to theology? Do any new theological insights emerge from the scientific study into the nature of the world? Conversely, can theology contribute at all in the scientific study of the universe and its origins? And how will theology's self-understanding be affected by the interaction?

When volumes and volumes have been written in a field, a single chapter cannot cover the entire gamut from primary scientific data to sophisticated theories.[1] But it *is* possible in brief span to present a representative sample of the *sorts of ways* in which scientific cosmology can contribute to theological conclusions, on the one hand, and on the other, of the ways scientific work stands in need of an interpretive framework of the sort that theology (and other metaphysical theories) offers. In order to provide a manageable first introduction to the field, I have chosen a representative thinker who serves as a clear example of each major category. As we move through the array of approaches, it will gradually become possible to identify the major paradigms in the field of religion/science, together with the concepts and assumptions unique

127

to each. Applying the criteria already worked out in previous chapters, one will begin to recognise the strengths and weaknesses of each major paradigm. The hope is that at the end of the day the reader will have a clearer sense of where the areas of fruitful interaction between theology and science lie, what pitfalls to avoid, and what conclusions that theologian may or may not draw from the scientific data. These results in turn will serve as guidelines for a more concrete theory of divine action in later chapters.[2]

NON-THEISTIC NON-MATERIALISM

One quick way to see 'how close and yet how far' scientific cosmology can come to supporting belief in God as Creator is to examine interpretations that come near to classical theism (or panentheism) yet diverge from it at the last minute. One good example of this phenomenon is Robert Wesson's *Cosmos and Metacosmos*,[3] which comes about as close as one can come to theism without actually asserting it. His reflection begins with Big Bang cosmology, the theory that the material universe or 'cosmos' began from a point about 15 billion years ago. According to him, it was generated from a matrix which he labels the 'Metacosmos' (pp. 8ff.). The universe as we observe it now is so utterly improbable that, Wesson argues, it appears as though it must have been designed for the development of intelligence. Yet one should not draw any theistic conclusions from the appearance of teleology, as for example advocates of the teleological proof have done (see Chapter 4 above). Instead, all one can say is that the 'Metacosmos' must contain *something* parallel to creativity, intelligence and purpose, albeit on a higher level than anything we know. The Metacosmos is not material, but it is not a spiritual being or force either. Hence one should conclude only that it is the foundation for the physical universe that we inhabit; it is outside both space and time but must have the capacity to create (produce?) a spatio-temporal order such as the one we observe. One thinks here of Kant's pure noumenal realm: it is not in space and time, nor do human categories such as causality apply to it; and yet noumenal things-in-themselves, Kant writes, are responsible for what we find around us – and in this sense are the causes for the phenomena we experience. Perhaps something like this concept is what Wesson intends.

Wesson's initial scepticism, his caution about drawing robust theological inferences directly from the cosmos, has much to be said for it. Wesson thinks there is reason to conclude that the universe points to *something outside itself*, and yet he is extremely cautious about developing a detailed metaphysics on the basis of physics. He is willing to say that the order in (and presumably behind) the universe is more fundamental than the material objects and regularities of behaviour that we actually observe. Yet beyond this minimal acknowledgement, he thinks, humans can hope to achieve only a partial explanation of the phenomena that we discover in the cosmos. The truths of physics are open-ended; every answer raises new questions, just as the evolution of life cannot be fully explained by neo-Darwinian theories of natural selection.[4] Ultimately, also *mind* is beyond complete comprehension; the relationship of mind and matter points us beyond any explanations we can find *within* the cosmos – and hence, again, toward the Metacosmos.

At the same time that Wesson begins cautiously and urges restraint, his work also provides a good example of a pattern one finds repeated again and again in the literature: after initial scepticism, the speculative engine gets going and a wide variety of bold metaphysical theories, models and metaphors emerge. Wesson moves ambitiously through the fields of cosmology, biology and philosophical anthropology, exposing the metaphysical puzzles to which they give rise, indicating his preferences for particular responses, and locating each within an overall cosmic movement that he discerns. In identifying this cosmic movement Wesson makes key use of the *anthropic principle*,[5] which stresses the significance of the evolution of the non-material side of the universe (intelligence or consciousness) for an understanding of the physical universe as a whole. To the classic anthropic argument he adds two principles. First, the origin of anything implies something grander (or, in Aristotelian-Thomistic-Cartesian terms, the cause of anything must have at least as much reality and perfection as the result), such that one must always speak of a movement from a higher source or cause to a 'lower' result. To this first, downward movement Wesson adds a second, upward one: the phenomenon of emergent properties means that life will always develop continually higher qualities. The result is a U-shaped movement, with the moment of creation as the lowest point: **U**.

Metaphysics covers the movement from the top left of the 'U' to the bottom or, in Wesson's terms, from the Metacosmos to the origin of the cosmos. Science then traces the movement up the right-hand side to where we currently stand (the emergence of intelligence or consciousness). Only religion, or art perhaps, can complete the upward movement of the ascending side of the 'U' at present. Admittedly, the scientific enterprise cannot forbid us from anticipating the 'U' in our minds. That anticipation even plays some role in science: aesthetic criteria such as parsimony and beauty or elegance influence theory selection, and religious beliefs about the final future of the cosmos can affect more short-term predictions and reflection within the natural sciences. Still, scientific study of the world simply cannot achieve real knowledge of the higher levels of existence that await our descendents as they move up the 'U' – except for the metaphysical principle that such a movement is inevitable.

USING SCIENCE TO CONVEY INSIGHTS ABOUT THE SPIRITUAL NATURE OF THE COSMOS

The writings of many recent cosmologists give rise to the strong impression that science is being used as a vehicle (or metaphor?) for conveying religious beliefs about spirituality and the rightful place of humanity within the cosmos. Angela Tilby's *Science and the Soul* is one good example of this category. Indeed, she makes no bones about it: 'What interests me . . . is the way in which contemporary science is acting as a catalyst to the transformation of religious ideas, particularly in spirituality.'[6] Such approaches should help warn theologians of the dangers of 'deriving' a doctrine of the God/world relation directly from science.

Of course, there is nothing wrong with looking to integrate the results of science and religion. Separating them forces us 'to either have the facts and renounce meaning, or cling to meaning in the absence of facts'.[7] Indeed, the dialogue between the two can be fruitful for both: metaphysical suggestions sometimes lead to empirically fruitful hypotheses (e.g. the Big Bang theory with its religious sources has done better against the evidence than its alternatives). Conversely, insights from the sciences can aid in religious reflection and perhaps even in the spiritual development of humanity (to say nothing of the quest for truth and knowledge!).

Unfortunately, though, Tilby does not really allow too wide a dialogue to develop with regard to theological positions. Her treatment of the types of theological imagery associated with major developments in science through the centuries[8] does not take a position in defence of a particular theological programme but remains purely descriptive: she *describes* how biblical imagery has changed in response to scientific developments. When one proceeds in this way, theological accounts come to function exclusively as metaphors – verbal accounts that may be of interest as expressions of humanity's spiritual nature but are neither true nor false of any state of affairs.[9] (Think of it as an updated version of Braithwaite's 'emotivist' theory of religious language, except that now the 'reduction' of language is not to the individual's emotional states but rather to her spiritual states.) Tilby's pluralism and her ability never to be dogmatic about theological positions are admirable, yet they are won at the cost of not taking any particular position within theology at all.

Positions of this sort are difficult to judge. Tilby, for example, enters into complex topics such as drawing parallels between anthropic approaches in cosmology, theoretical developments in physics and themes in Christian theology.[10] She is also right to challenge easy theological appropriations of the anthropic principle, those that find in universal evolution clear evidence of God as its guiding principle, arguing instead that one could just as easily conclude that evil is a necessary component of creation.[11] She is also right to insist that the anthropic principle, if it holds, should not be equated with *anthropocentrism*, that is, an account of the universe that places *humanity* at its centre as its most important component. Depending on what metaphysics one brings to them, the so-called anthropic arguments in cosmology can be used to support belief in a higher being (theism), or in the inherently spiritual nature of the unfolding cosmos (pantheism)[12] – *or* they can be interpreted minimalistically, as statements of what is required (for example) to resolve the wave function but with no broader metaphysical significance than that. Tilby is thus right to caution against straight-line arguments from physics to metaphysics.[13] Talk of the importance of the observer in cosmology *may* be good news for the doctrine of creation, even leading one to posit God in order to provide the background stability for the universe or to guide its development. Likewise, the process of evolution *may* eventually provide grounds for talk of divine guidance (*if* theologians are able to develop a scientifically sophisticated account of the need for a theistic component in evolutionary theory). But it is over-hasty to claim that such entailments *currently* exist in physics or biology.

Indeed, on some points even *more* caution is required than Tilby seems to think. Physics may well discover the mechanisms that helped to shape the development of the universe and of life within it. For example, it may find in chaos theory resources for explaining how life can work 'against' the second law of thermodynamics and continue in the direction of increasing complexity. But even breakthroughs of this sort only *describe* what has happened; they do not tell *why* life should start developing in the first place. Another example: the question of eschatology[14] must continue to trouble science/theology dialogues, since the best evidence suggests that the universe will be uninhabitable for humans some 30–40 billion years before its end (assuming the 'Big Crunch' cosmology). To move too quickly towards the integration

131

of science and religion is to risk doing injustice to the remaining tensions.

Where conflicts between theology and scientific conclusions do arise, many authors are inclined to shift the focus away from the topic of which theories are true. In *Science and the Soul*, for example, Tilby takes what she saves on science and metaphysics and spends it freely in the area of science and spirituality. Like many, she finds in the human sense of a spiritual dimension to the universe a comforting resting place when direct inferences from science to religious *beliefs* seem blocked. One other common response makes its appearance in this book: the stress on the nobility of *the human effort to understand* rather than the advocacy of any particular position. Typical of the preference to focus on the fact 'that we can ask the question of our origin at all' rather than on 'what the ultimate origin and telos might be' is her endorsement of the well-known phrase from Steven Weinberg: 'The effort to understand the universe is one of the very few things that lifts human life a little above the level of farce, and gives it some of the grace of tragedy.' It may be that theologians will someday have to limit their concern with science to, say, the sense of God's presence in nature as a whole, or to the sense of a spiritual connection of humanity with the universe. But, at least for the time being, there remains a whole batch of fascinating and vitally important metaphysical issues to be raised, and perhaps settled, in response to scientific developments in our century.

SCIENCE USURPS THE ROLE OF THEOLOGY

The career of the physicist Frank Tipler serves as an interesting example of another model for relating science and theology. As Tipler worked on the 'origins' question in physical cosmology, he discovered that the role of the human observer might be much more significant than physics had assumed. The fundamental values of the universe are slanted towards the eventual emergence of higher life forms in such an improbable manner; it is as if the emergence of life were a necessary feature of the universe. Together with John Barrow, Tipler wrote the now-classic *The Anthropic Cosmological Principle*.[15] The book helped bring home to a broad reader-ship that the values of a large number of fundamental physical constants lie within an extremely narrow range – the only range that would allow for the emergence of life. The observation of the a priori improbability that all of these values would line up in precisely this manner so as to allow for intelligent life is today called the *weak anthropic principle*.

(Of course, others have responded that the probability of *anything* is

100 per cent once it has occurred. The a priori probability of my winning the jackpot in the lottery is, say, one in 7,000,000. However, *given* the knowledge that I *have* just won the lottery, the probability of my winning it jumps immediately to 100 per cent. Similarly, given your knowledge that you exist, you cannot judge the probability of life existing to be low. Of course, if you were looking at a whole line of possible universes and did not know which one would be actualised, then you would have to judge the probability of intelligent life arising as rather miniscule. But if you dwell in a universe in which there is life – and obviously, since you dwell in it, it is such a universe! – then the probability that universal evolution would bring forth life immediately jumps to 100 per cent. Only if you considered a universe prior to knowing its fundamental constants, and prior to knowing whether life would arise in it, could you speak of life being unlikely. But no thinker in a universe could ever be in that position!)

Barrow and Tipler's *strong anthropic principle* is a rather harder pill to swallow. Based on work by John Wheeler and others, it involves the assertion that the universe *had* to bring forth life, in order (among other things) to resolve the quantum probability functions into a given macro-physical state. Under one interpretation of quantum theory, a photon of light travelling for millions of light years through the universe is strictly speaking only a probability function until the moment that it strikes your eye at night; at that moment it is retroactively resolved, back through those millions of years, into this particular photon. According to Barrow and Tipler's argument, the arising of life had the same sort of necessity to it: there would not have been an actual universe without the eventual emergence of an observer who would retroactively resolve the quantum states into the macrophysical universe that we know.

Although within physics the role of anthropic principles has not yet been resolved, it is a debate to which theologians should pay close attention. Just as it is remarkable that we would have hard evidence of the universe's origins, so also it is amazing that there should be *physical* reason to think that the presence of intelligent observers might be either essential to the universe or part of its fate. As Peacocke writes, 'The most striking feature of the universe is the one that is so obvious that we often overlook it – namely the fact that we are here to ask questions about it at all.'[16] Mistakes lie close at hand if one seeks to draw all-too direct inferences to divine intentions. Still, one cannot ignore the fact that there is some physical reason to think that the arising of intelligent life is something more than coincidence and that we are therefore more than a contingent product of random universal evolution.

Tipler's more recent book, *The Physics of Immortality*, raises another set of questions and offers another model of science/theology interaction.[17] It is an ambitious undertaking: his goal is to show that physics and the Christian tradition lead to some fundamentally similar conclusions. If *The Cosmic Anthropic Principle* was present tense, describing the significance of present-day observers in the history of the universe up to this point, *The Physics of Immortality* is future-oriented. It extrapolates (a version of) the anthropic principle into the final days of the universe. The same universe that had to lead to the evolution of conscious beings must also be one in which they are destined to grow in knowledge and control. Intelligence will not be wiped out at some future point. Rather, the book aims to show, the most likely hypothesis is that those intelligent beings who are our descendents will still be there at the universe's end. Of course, they will not be carbon-based intelligent beings, since the conditions for life as we know it will hold for only a relatively few billion years; somewhere around the universe's 60 billionth birthday the class of suns that we rely on to sustain life will have burned out. Tipler argues against the Steady State theory in cosmology, according to which the universe will end in an eternity of near motionless entropy, advocating instead a final implosion of the universe as the better theory. As the 100 billion or so galaxies draw closer together again and distances decrease, it will be possible, he argues, for a non-carbon-based (perhaps silicon-based?) life form to gain more and more knowledge of the universe and to communicate throughout it – until, at the very last instant before the Crunch, these intelligent descendants of the human race will know everything there is to know about the universe. They will be omniscient or divine, and the universe will be fully self-knowing.

A briefer version of Tipler's immortality theory can be found in 'The Omega point theory: A model of an evolving God'.[18] Here Tipler provides what he claims is a physically consistent account of God. The view is panentheistic, he says, because 'God' – in the sense described in the previous paragraph – becomes the descendent of all life, though it (they?) is also more than what came before. God is the moral and intellectual force(s) in the universe, manifesting both the capacity of recollection and the ultimate intellectual power at the close of the universe. Consequently, Tipler's 'evolving' God is immanent on the one hand and transcendent, omnipotent and omniscient on the other – at least at the final singularity that his theory predicts. Because it possesses the power to know all things, and in some sense to control all things (though surely it doesn't control the final destiny of the universe!), this Intelligence provides the universe with a sort of ultimate meaning.

Tipler is a strong spokesperson for the view that one can begin and end with physical science without advocating a reduction of all entities and properties *to* science. The arguments for his view, he maintains, are founded on physics alone, although his sources of inspiration include the New Testament and some theologians, most notably Wolfhart Pannenberg. The theory yields testable predictions – well, at least they are falsifiable in the very long run. Yet it emphatically does not reduce everything to physical particles and forces, since its focus is on the final results of the universe's evolution and on God-like properties such as intelligence. Like a number of other contemporary options, Tipler's physical theory shows that one can listen closely to the physical sciences and be guided by their results without having to accept a reduction of all things to physical properties.

Like other physicists who try to work their way *from* the science *toward* religious conclusions, however, Tipler cannot be faulted for being uncontroversial. It would be fair to say that his 'physics of immortality' has not won immediate acceptance within the scientific community. Indeed, readers in the field may be struck by the radical disagreements one finds among the more speculative interpretations of recent physical results. Such divergence of opinion brings home powerfully the difference between the sense in which empirically based physical theories are speculative (only minimally!) and the sense in which *interpretations of* (or arguments over the implications of) physical theories are. Different authors speculating in the latter fashion come to massively different conclusions.[19]

Nor for that matter can Tipler be faulted for his orthodoxy. His is a future God, not one who exists currently. As we saw, God will exist fully when life within the universe reaches a certain level of development, namely when *information processing* becomes actually infinite. When intelligent life has become omniscient, then we can and must say that God is complete. Recall that Tipler set as his goal to defend a viable *physical hypothesis* which, if true, would describe a state of affairs that is at least consistent with – and in this sense can be understood as the realisation of – biblical eschatology. On his model of science/religion interaction, it appears, to advocate theological theories about the universe that go beyond what can be 'parsed' scientifically would be irrational. Instead, one does one's science first and then tells the theologians the sense in which their beliefs *might* be empirically grounded. Whatever cannot be grounded in this way is presumably a free flight of fancy.

I suggest that this sort of linear approach and final control by physics represents a straitjacket on theological reflection that is artificially tight.[20] An adequate doctrine of God and of divine activity in the world cannot be obtained under the tutelage of physics alone. Three critical reservations will have to suffice. One is struck, for one, by the unnatural way in which Tipler applies New Testament statements about immortality and eternality directly to his physical theories. There is no acknowledgement of a difference of kind, no sense of the otherness of (at least some parts of) theology to scientific cosmology. Also – presumably as a result – Tipler insists that any conception of a transcendent God who is 'other' to the world is superfluous. 'Divine' omniscience and omnipresence refer to future states of the physical universe, and eternality means living as long as there is a physical universe around to live *in*. Of course, there are reasons – good scientific reasons, if not traditionally good theological ones – for limiting the referent of all our terms to the universe as it is now or potentially one day will be. The downside, however, is that what might possibly be beyond the universe – a state of being prior to or following the universe's physical history – thereby becomes ineffable. Although Tipler allows for a transcendent Ground (the Omega point), he lacks the vocabulary to give it a conceptual explication. This is a cost that theologians should not be willing to pay.

Finally, where Tipler *does* use theological terms, their substance often seems to have been evacuated. Tipler wishes, for example, to call his position panentheism: God is reducible to but not describable by the history of life taken as a whole. Does 'not describable' in this phrase map adequately (even approximately) onto the transcendent pole of panentheism as we have been studying it? Note that Tipler's 'indescribability' represents a limitation of *knowledge* more than it does the claim that God is something more than all the intelligence in the physical universe. The ontological claim that God includes but transcends the physical universe, which is a (the?) defining characteristic of panentheism, is insufficiently pronounced in Tipler's positions. It thus seems more accurate to describe Tipler's position as a kind of pantheism, a sort of scientifically updated Spinozism: 'God' (for Spinoza, *deus siva natura*) is another word for the universe taken as a whole, including both its physical and mental properties, or what Spinoza called the 'attributes' of thought and extension.

SCIENCE ITSELF CHALLENGES REDUCTIONISM AND SUPPORTS THEISTIC DESIGN

Another group of thinkers draws major inferences for theology directly from scientific cosmology, albeit without, like Tipler, trying to make the science itself do the work formerly assigned to theologians. One of the best known is Paul Davies, who has argued over a number of years that physical reductionism is counterindicated by an adequate understanding of current scientific cosmology. Because of Davies' prominence, and because he offers something like a science-based teleological proof, he may serve as a paradigm example of this group of thinkers.

Paul Davies first attracted the attention of theologians with his *God and the New Physics* in 1983.[21] This book was in some ways the theist's response to an earlier batch of books that had emphasised parallels between contemporary physics and the Eastern religions, particularly Buddhism and Taoism.[22] In a very accessible manner Davies was able to show that recent developments in physics and cosmology have reduced the distance between theistic belief and physics. Instead of trying to make physics itself religious, he contented himself with points of contact, with relative probabilities, and with metaphors allowable in the contemporary context that are congenial to theism. A number of Davies' controversial claims in that early book seem less controversial now: a universe arising from a Big Bang *is* more congenial to the Christian doctrine of creation than is an Aristotelian universe of eternal matter, and the relativistic (Einsteinian) move to convertable matter and energy is more congenial to spiritual forcefields than is a physics based on particles of matter, in which energy is merely a secondary appearance.[23]

In later books, Davies began to emphasise more strongly the order in the universe and the strength of inferences from it to the existence of design. For example, in *Superforce*, as in many of his later books, Davies concentrated on phrasing the physical theories in layperson's terms, suggesting religious connections in only a most cautious way. Only in the final pages does he come, carefully, to his own sense of the compatibility – or stronger – of the physics and theism:

> Should we conclude that the universe is a product of design? The new physics and the new cosmology hold out a tantalizing promise: that we might be able to explain how all the physical structures in the universe have come to exist, automatically, as a result of natural processes. We should then no longer have need for a Creator in the traditional sense . . . Nevertheless, though science may explain the world, we still have to explain science. The laws which enable the universe to come into being spontaneously seem themselves to be the product of design . . . [T]he universe must have purpose, and the evidence of modern physics suggests strongly to me that the purpose includes us.[24]

Davies has seen that theological assertions cannot be derived directly from modern science. The more urgent task for theology, he recognises, is to address, and if possible to refute, those interpretations of science that would eliminate a role for God from the start. Only if theologians are successful at this task will the door be opened for a theological interpretation of natural science. Thus in *The Cosmic Blueprint*[25] Davies makes the case against reductionist accounts of physical law, in so far as they cannot satisfactorily explain the tendencies within the macroscopic world toward progressive organisation.[26] Evolution is a process that leads to *increased* organisation; as such it runs *against* the second law of thermodynamics, which states that closed physical systems will show an inevitable increase in entropy with time. (If you knew the oxygen in a closed room was evenly distributed at one point in time but concentrated in one corner at another, you will know, assuming no other outside disturbances, which state came first.) In the evolution of life the 'arrow of time' seems to point in the opposite direction, away from increasing entropy; life forms become *more* complex with time. Hence, it seems, we have evidence that a reduction to physical explanations is inadequate for the biological data.

Davies draws several conclusions from this evidence, which we might reconstruct in terms of increasing ambitiousness.

1. It seems impossible to explain the biological phenomena using physical laws alone; they require their own set of (irreducible) laws.
2. Life is better seen as an 'emergent property'; it is not present at lower levels of complexity but 'emerges' only in higher forms, which must therefore be grasped in terms of their own principles (more on this below).
3. To explain the increasing organisation in the biological sphere, we need 'theories of organization', broad theories that account not only for the tendency toward increased organisation but can also explain the actual forms in which they are organised. For example, the explanatory task in the case of humans is to account for phenomena such as consciousness. Mental properties such as freedom and intention, clearly being something more than deterministic physical properties, need their own principles of explanation.
4. Science may well remain the one reliable path to knowledge, at least of the physical universe. Nonetheless, a study of these broader theories of organization, Davies claims, points toward the existence of an overall purpose within or behind the universe.

This last contention makes Davies an advocate of some form of 'teleological' argument for the existence of God. According to this view, the universe is founded on some creative principle or principles, even if the use of mystical or theological explanations is less conducive to knowledge of these principles than is science.

What is to be said of Davies' more ambitious claims? (1) and (2) are key insights that should be basic parts of every theologian's repertoire today. Theologically, (1) is a direct implication of 'In the beginning God created the heavens and the earth.' It means that the world cannot be explained in physical terms alone but requires 'higher' explanatory principles, without yet specifying *how far* they must diverge from physical principles. Likewise, (2) asserts that there are discontinuities in creation: animals have qualities not found in the physical world, humans qualities not found in other higher primates. Traditionally, theologians spoke of the human soul; today we might speak of features of human existence such as rationality, morality, responsibility, choice or freedom – and the religious attributes such as love, sin and eternality. Put in theological terms, (3) conveys something of the *order* with which God created the universe and which is still expressed within it – and perhaps also the ethical beliefs which anticipate the final order that God will establish when he brings all creation to its final culmination or perfection (*telos*). It is basic to the scientific task to look for such signs of order in the world, especially order expressed as mathematical regularity or law-likeness, just as it is basic to theology to explain the existence of order – both the search for it and the fact that we do find it – in terms of its divine source.

But what of (4), the design argument? Consider for a moment Davies' recent book, *The Mind of God*. Davies is (sometimes) cautious; the book is not, as the title might indicate, an attempt to read off the mind of God from the physical universe. Nonetheless, Davies does believe that religious, and specifically theistic, conclusions can be drawn from the evidence provided by the scientific study of the universe. Indeed, he thinks conscious design is the *more likely* conclusion to draw:

> The central theme that I have explored in this book is that, through science, we human beings are able to grasp at least some of nature's secrets . . . What is Man that we might be party to such privilege? I cannot believe that our existence in this universe is a mere quirk of fate, an accident of history, an incidental blip in the great cosmic drama. Our involvement is too intimate. The physical species *Homo* may count for nothing, but the existence of mind in

some organism on some planet in the universe is surely a fact of fundamental significance. Through conscious beings the universe has generated self-awareness. This can be no trivial detail, no minor byproduct of mindless, purposeless forces. We are truly meant to be here.[27]

Now how far is this argument grounded in the scientific results themselves and how much belongs to the level of (metaphysical) interpretation? As we saw in discussing the theistic proofs in Chapter 4, the theologian cannot help but look for signs of divine order in the cosmos – for signs that 'we are truly meant to be here', signs that serve for her as a sort of confirmation of her belief. But recall the example of the lottery above. If I know that the outsider Thrown-a-Curve has beaten the odds and won the Kentucky Derby, I can look back and find signs of his strengths – his lineage, his training, his performances at previous races – which hinted in advance at his upset win. But this is hindsight, since it compiles the evidence in light of the later knowledge that he *had* in fact won. Before the Kentucky Derby, the odds of his winning were not decisive. The same logic applies to the teleological argument. If one observed the Big Bang, knowing the a priori odds against life ever arising in such a place, and *then* one watched the gradual evolution of planets and of life on this planet (and perhaps others) up to the present, one might be hard-pressed to avoid the conclusion of divine guidance; the design argument would stand. But if one *begins* with the knowledge that life has arisen (similar to beginning with the knowledge that Thrown-a-Curve has already won the Derby), then appeals to the design argument are less compelling.[28]

Theists should also grant that other possible explanations of what we observe also decrease the strength of the argument. If the 'multiple universes' theory is true, and there really exist a large number of universes, each representing an array of fundamental constants, then there is nothing unlikely at all about our existing in this universe, since (according to this hypothesis) a universe with life did not arise *instead* of all the others, but rather *along with* the other uninhabited universes. Likewise, finding a large variety of life forms in this universe, including other intelligent life forms – and very recent new evidence of planets similarly situated to our own has significantly increased the possibility that this is the case – would also make our own emergence seem less 'strange' and unlikely, and hence less in need of an explanation, theistic or otherwise. (I suggest that faith works the same way: if one already has faith that God is guiding universal history, it is easy to find signs of that guidance in this or that event or pattern. But if one tries to make the case for divine interventions in a way that would convince a neutral observer who does not already share this faith – not to mention one who is disinclined to believe it! – one finds the task rather more difficult.)

Hence I suggest that the conclusions to be drawn about the teleological argument should be closer to the cautious inferences that Robert Jastrow makes in *God and the Astronomers*.[29] Jastrow shows how recent developments in cosmology support the concept of a universe with a beginning, but he is an agnostic about theological matters. It does not seem that a decisive case can be made from what we know of the evolution of the universe and our place in it alone. Significantly, Jastrow quotes the famous anecdote about Einstein: 'When Einstein came to New York in 1921 a rabbi sent him a telegram asking, "Do you believe in God?" and Einstein replied, "I believe in Spinoza's God, who reveals himself in the orderly harmony of what exists".'[30] Einstein's answer parallels Davies' stage (3), since it appeals to higher principles of order that must be introduced to explain what we find around us. Einstein famously used the word 'God' to express the types of order he felt he had to presuppose in explaining the universe; but equally famously he refused to supply 'God' with any more content than that presupposed order itself. When we move beyond (3), we move into the realm of metaphysics (or theology proper), formulating broader explanatory theories that attempt to make sense of experience as a whole – not only experiences of the physical world but also cultural, moral and existential experience. Here the database is so broad that no decisive answers are forthcoming, even though one can still give reasons for and against particular metaphysical answers.

Note that our discovery of the need for caution before developing theologies of design neither gives the theologian *carte blanche* to speak of God's action in and behind the world nor counts against the possibility of such a theory. If anything, it underscores the importance of reflection beyond physics, since it shows how important questions are raised that cannot be answered in that context. Nonetheless, such theological reflection about the physical world is not unconstrained. Theologians who write about nature make, and must therefore defend, explanatory claims at the broadest level. As Peacocke argues, 'from the existence of the kind of universe we actually have, considered in the light of the natural sciences, we do infer the existence of a creator God as the best explanation of all-that-is'.[31] In questioning the adequacy of teleological arguments I have questioned the strength of this inference. More accurate would be to say that we discover, and seek to show, the continuing adequacy of belief in a Creator God as an overarching explanation for the existence of the universe as science has come to understand it. Still, Peacocke is right to insist that a theology of creation cannot preserve its intellectual integrity without entering into the competition for the best explanation of the physical world.

THEOLOGY HAS THE INTERPRETIVE AUTHORITY OVER SCIENCE

It is tempting for the theologian to set down the conditions on science in advance. 'Here is what we know about nature, here is how it must be,' she specifies in advance, 'now let the scientists go out and learn what they may. If they come up with other conclusions, however, we will reject their findings, since we will know that they have made atheistic or naturalistic assumptions that are unacceptable to us.' Such positions represent a popular paradigm for the science/theology relationship.

Ted Peters is a theologian at the Graduate Theological Union in Berkeley who has been one of the major voices in the science/theology dialogue. Not only has he reflected theologically on major scientific developments,[32] but he has also been involved in a firsthand way with actual projects, for instance serving as principal investigator on the ethics side of the human genome project.[33] Both as editor of *Cosmos as Creation*[34] and as the author of an influential typology of positions in theology and science,[35] he has helped to make clear the dangers and opportunities for theology in its discussion with science.

In 'On creating the cosmos'[36] Peters argues that enough progress has been made to relax the separation between the neutral language of science, which is concerned with 'how' questions, and the language of theology, which raises 'why' questions (p. 275). Peters advocates an active search for 'hypothetical consonances' – harmonies between these disciplines (p. 274). The remainder of the article explores the Christian concepts of *creatio continua* and *creatio ex nihilo* as they relate to thermodynamics and Big Bang theory. Peters feels that these views are not exclusive and that God has/does implement both (i.e. creation of a dependent but separate world). He finds evidence of both in scientific cosmology.

I am in many respects in agreement with Peters. However, he does draw attention to a certain danger in this type of discussion. The comparison between the two fields becomes suspect if scientific assertions are viewed as 'flexible' but theological assertions are viewed as immune from radical revision. In this way the theologian assumes her foundation as given, and picks and chooses which parts of science confirm her perspective but disposes of the problematical concepts. Clearly, a more symmetrical approach to the problems would help alleviate any charge of bias on the part of theologians.

FAITH 'DISCOVERS' BELIEF IN THE CREATOR AS THE BEST ANSWER

Some make a stronger claim on behalf of theism: the claim that faith in the 'making of heaven and earth' by a Creator is the metaphysical foundation that (alone) can ground both science and philosophy. Consider, for example, the argument in *Cosmos and Creator* by Stanley Jakti.[37] Jakti maintains that there is no incompatibility between science and religion when they are mediated by faith in a Creator God; rather, on this basis (and only on it) one realises that both are legitimate and that the future progress of humanity requires both.

Like many religious 'compatibilists', however, Jakti ends up challenging science. He urges, for example, a greater scepticism about scientific knowledge claims in order to correct for the overemphasis on science in our day: 'Thinking when done by scientists is not necessarily the source of truth, scientific or otherwise. More likely than not, it is suggestive of presuppositions whose truth rests with the scientist's philosophy and not with his science.'[38] Because of the role of philosophical assumptions in allegedly pure scientific arguments, we should be much more cautious about appeals to scientific and mathematical 'proof' as the standard for knowledge (pp. 102f.).

The reason that science and theology have often been viewed as opponents is to be found in mistakes that philosophers have made, claims Jakti. Hence the solution must lie in correcting these philosophical errors:

> The cure of modern philosophy will perhaps come from a reflection on its persistent malaise. Modern philosophy is caught, more than was the case with ancient Greek philosophy, in a hapless oscillation between two extremes: the mirage of absolute certainty and utter skepticism. Philosophy transcended this predicament only when during the Middle Ages a widespread adherence to the doctrine of creation kept philosophy on the ground of metaphysical realism, the only safe ground between the abysses of absolute certainty bordering on tautology and a no-certainty-at-all provoking despair. Not until that ground is recovered will there be an unequivocal surrender to the reality of the cosmos, the indispensable stepping

stone for a rational recognition of the existence of the Creator. Such is the only solid basis for looking at existence not as a global trap and for mustering the resolve to view the earth as a genuine home. But just as that ground was not gained in the Middle Ages, the age of faith, without the guidance of faith, it will not be regained without first recovering that faith. *In a very crucial sense, one must first say Creator in order to say Cosmos.*[39]

Jakti is right to focus the science/theology discussion on underlying views of the nature of knowledge (epistemology) and 'what is' (metaphysics), for if these underlying (or hidden!) views *exclude* faith or a religious answer, then *no* analysis of scientific results can be compatible with theology. (Conversely, if one assumes that reality is fundamentally religious, it will be no surprise when one then 'discovers' that one was right all along!). This is why one cannot do theology today without an adequate consideration of how theological truth claims relate to knowledge claims in other areas.

At the same time, I fear that this sort of approach makes the theologian's task too easy, by which I mean easier than the evidence warrants. Theological work of this kind does not need, for example, to confront 'hostile' or dangerous conclusions in science, since 'the eyes of faith' can always eliminate any incompatibility. But are not some data or theories about the physical world in fact *more difficult* to explain in theological terms than others? For example, the fact that our solar system will be unable to sustain life as we know it in the finite future is more difficult to account for theologically than the fact that the universe apparently began with a bang some 11–15 billion years ago. Similarly, it is easier to explain the existence of rational-moral beings who evidence religious behaviours by positing an all-good, all-powerful God than it is to explain theistically why human life would have developed through a process that entailed the extinction of countless species and the suffering deaths of billions of organisms.

Examples such as these suggest that the task is not just a matter of choosing a metaphysic that allows one to view the universe theologically as God's creation. The real challenge is to study which metaphysical views are suggested by the various scientific results and which are *counter*indicated by them. It is well known that Hoyle advanced Steady State cosmology because he recognised that an initial singularity, if widely accepted, would provide powerful evidence for theists *and he strongly believed that theism was mistaken.* It may be that an approach to the world in purely physical terms is inadequate to the religious dimension of human existence. But its inadequacy must be *shown* – as thinkers like Peacocke, Russell and Pannenberg have attempted to do in various ways – rather than presupposed at the outset.

HOLISM WITHOUT TRANSCENDENCE

The early reactions to relativity theory and quantum mechanics by physicists were more philosophical than religious. In now classic works, Sir Arthur Eddington, Werner Heisenberg, Carl von Weizsäcker and others began to analyse the philosophical implications of the new physics for which they were responsible.[40] It is a tradition that has continued down to the present day.[41] Among these works, David Bohm's was one of the first to have a clearly religious ring to it.

First Bohm challenged the determinism that had dominated modern physics since Newton. With the developments in quantum physics, he saw, chance now had to be admitted as a key component in even our best physical theories:

> First of all, we point out that *if* there are an unlimited number of kinds of things in nature, no system of purely determinate law can ever attain a perfect validity. For every such system works only with a finite number of kinds of things, and thus necessarily leaves out of account an infinity of factors, both in the substructure of the basic entities entering into the system of law in question and the general environment in which these entities exist. And since these factors possess some degree of autonomy (e.g. *stability* or *existence*), one may conclude . . . that the things that are left out of any such system of theory are in general undergoing some kind of a random fluctuation. Hence, the determinations of any purely causal theory are always subject to random disturbances, arising from chance fluctuations in entities, existing outside the context treated by the theory in question. It thus becomes clear why chance is an essential aspect of any real process and why any particular set of causal laws will provide only a partial and one-sided treatment of this process, which has to be corrected by taking chance into account.[42]

Bohm next realised that the ontology of independently existing objects – or even that of individual particles, the core concept of Newtonian physics – needed re-examination. Mechanistic philosophies of this sort, he argued, are arbitrary and unsupportable. The *number* of things we take there to be depends, *inter alia*, upon the purposes for which theories are constructed and how we measure the world (this claim has been further substantiated by recent work in mereology, the set-theoretical calculus of parts and wholes developed by Lezniewski). What 'really' exists, then, is a differentiated whole; any parts that we divide out must therefore be viewed as interdependent. Bohm thus stressed the innate tension between 'autonomy' and 'reciprocal relationship' in the concept of a thing:

> The notion of a 'thing' is thus seen to be an abstraction, in which it is conceptually separated from its infinite background and sub-structure. Actually, however, a thing does not and could not exist apart from the context from which it has thus been conceptually abstracted. And therefore the world is not made by putting together the various 'things' in it, but, rather, these things are only approximately what we find on analysis in certain contexts and under suitable conditions.[43]

At the conclusion of this work, Bohm makes clear the theological stakes of his interpretation of the implications of quantum theory: 'In conclusion, a consistent conception of what we mean by the absolute side of nature can be obtained if we start by considering *the infinite totality of matter in the process of becoming* as the basic reality.' In the physical world, 'relativity is absolute'; as a result, 'the essential character of scientific research is, then, that it moves towards the absolute by studying the relative, in its inexhaustible multiplicity and diversity'.[44]

This suggestion became the core concept of Bohm's influential *Wholeness and Implicate Order*.[45] Bohm sees reality as analogous to a holographic image in a state of constant flux: the 'part' is built up from an imperfect notion of the 'whole'. Classical or Newtonian 'movement' is replaced by the concept of 'holomovement'.[46] As Kevin Sharpe summarises it, the view rests on five central metaphysical assertions: (1) reality is characterised by infinite depth and (2) relation; (3) motion is primary, not stability; (4) this motion is *creative* – reality is always transforming itself; (5) reality divides into systems, levels or hierarchies.[47] Perhaps as a result of his religious influences – Bohm was raised Jewish but became fascinated with Eastern thought, particularly Krishnamurti – Bohm holds that it is only the illusion of separation that distorts our conception of the world. Consciousness, correctly understood, is an indivisible whole.[48]

Sharpe tries to argue that Bohm's reflections on quantum theory could be appropriated by Christian theologians. Under this view, God would be both transcendent *and* immanent (overtones of panentheism?), and

God *could* be personal.[49] But Bohm's vision of the world, it seems to me, has much more in common with Spinoza (who as I recall was never mentioned in Sharpe's book) than with any version of theism, including panentheism. Bohm's personal philosophy seems summed up more accurately as an extension of pantheistic mysticism.

The specific doctrine of God and the version of the God/world relation that would be derived from this interpretation of quantum physics may be only minimally helpful to the classical theologian. But Bohm's work on the part/whole relation remains an important guiding principle for theology. Bohm – and Bernard d'Espagnat after him[50] – argue from the primacy of fields over particles in quantum field theory to a general (metaphysical) primacy of the whole over the part. In Chapter 8 I will suggest that something like this primacy of whole over part should guide our understanding of emergent levels of explanation (and presumably of reality) as we move from physics to biology and from biology to psychology. Ultimately, just as the holism of mental states supervenes on the complexity of brain states, the level of the divine stands above, encompasses and sums together all finite reality. Bohm may resist an emergence theory of the God/world relationship. Still, his work on the primacy of the whole may well prove fruitful to theologians working on the God/world relationship 'in light of science' in ways that Bohm himself did not imagine.

THE DESIRABILITY OF MAINTAINING A MYSTICAL (PANTHEISTIC) RESPONSE TO THE WORLD

A number of thinkers begin from the assumption that (something like) Kant's critique of metaphysics has made it impossible to do theology, that is to formulate beliefs about the nature of a transcendent being. At the same time, a world-view formed by science alone would be barren and devoid of meaning; and, they argue, many scientists *do* retain some sort of spiritual response to the world. Could there be some way to preserve the *function* that religion has traditionally fulfilled – providing a 'sacred canopy', a framework that protects one from the sense of *anomie* and chaos in the physical world[51] – without getting into the metaphysical morasses that belief in God seems to entail?

Edward Harrison serves as a good example of this (widespread) approach. Harrison approaches cosmology with a strong dose of Kantianism. He holds that the Universe (that is, the universe in itself) is fundamentally unknowable, and that all interpretations that we put upon the Universe are founded in particular social/cultural contexts. Harrison endorses 'learned ignorance'[52] as the rational alternative to positive articulations of metaphysical knowledge. What remains when we are hard-headed enough to dispense with all metaphysics? Only a pantheistic/mystical appreciation of the Universe, supplying meaning in an otherwise barren universe:

> We feel an urge to believe in God, and this desire derives not from arguments of pure reason. Articulated in holy lexicology, hallowed by tradition, the urge is admittedly irrational within the context of the physical universe. Rationalists resist the urge on the grounds that it emerges from the jungle of misguiding elements in our cultural heritage. Many rationalists and most agnostics realise that absence of proof is not proof of absence, and therefore they take an occasional interest in arguments claiming to show that we have the necessary and sufficient reasons for believing in the existence of a supreme being.[53]

Harrison argues that humans cheapen their existence by moving away from religious belief. Such a move is more an expression of prejudice based on the success of science than a response to argument. Moreover, the consequences of atheism are negative, since atheism goes against our very nature:

> One may legitimately argue that as a result of rejecting the notion of gods, our views concerning reality have become pallid and inane, and our views concerning life itself devoid of satisfying social and personal meaning. Our cultural heritage impels us to believe in God or something similarly all-inclusive and inconceivable, for it has stolen from the phenomenal world the very elements essential for a life of significance and given these elements to the gods who have the function of sharing them with us.
>
> One has only to imagine the home with its own house god or hearth goddess, who emanates an ambience of warmth and friendliness, wards off danger, and is acknowledged by libations and flowers, for the home to seem more secure and restful. This fanciful pagan illustration shows how deep within us is a strong urge to live in fellowship with the gods, and when we deny their existence and live in a godless universe, we are left with a residue of unfulfilled yearning.[54]

When one surveys the world-views that have oriented human lives up to and including the present – and the bulk of Harrison's book is devoted to a historical examination of the various conceptual universes humans have lived under – one realises that the scientific world-view cannot and should not be humanity's final resting point. For one thing, scientists do not just proceed dispassionately, as the myth would have it, in the attempt to falsify their own theories and to advance 'knowledge in general'; they do not really exemplify the model of science advocated by the influential philosopher of science, Karl Popper.[55] If dispassionate objectivity is a myth in science, how much more so must it be a chimera at the level of one's world-view as a whole? Furthermore, the scientific way of thinking and approaching the world simply does not supply an adequate overarching framework of belief for humans.

What kind of religious world-view, then, would serve the function Harrison advocates? Significantly, when he describes the kind of religious belief he thinks is needed, he comes closest to pantheism:

> If we can think that all is far from known, and God is perhaps the Universe, then without further intellectual commitment we avoid the dreariness of atheism. By equating God and the Universe we give back to the world what long ago was taken away. The world we live in with our thoughts, passions, delights, and whatever stirs the mortal frame must surely take on a deeper meaning. Songs are more than longitudinal sound vibrations, sunsets more than transverse electromagnetic oscillations, inspirations more than the discharge of neurons, all touched with a mystery that deepens the more we contemplate and seek to understand.[56]

This passage, like much of the literature in the field, reflects the assumption that pantheism is the default position. It allegedly requires a smaller dose of metaphysics and fewer specific tenets; yet at the same time it avoids atheism, which Harrison maintains (correctly, I think) does not do justice to the religious side of human nature. Those who advocate this position seem to have as their main goal to preserve a place for human spirituality against interpretations of the world and science that would eliminate or marginalise it.

Theologians have several possible responses to 'default pantheism'. One can argue that theism is in fact no more complicated (that is, a priori unlikely) than is pantheism. Since we have immediate knowledge of what it is to be an agent acting in and on the world, it is a simple move to posit a personal source of the world; and have not the critics of theism, who charge it with the arbitrary projection of human categories onto the universe, used precisely this simplicity as an accusation *against* theism? I do not find this response fully convincing, however, because pantheism is indeed a metaphysically minimalist position. As much an *attitude toward the world* as a metaphysical position, it can be limited to the belief in some ultimate Real underlying the phenomenal world (John Hick).[57] Or it might connote merely the sense of a spiritual dimension in the world, positing no additional beings or entities at all but merely suggesting that one interpret the world *as* mystical or spiritual in some sense.

The second response is for theologians to challenge the suspicion of metaphysics in contemporary thinkers such as Harrison and Tilby. Two facets of this response have already occupied us above, as we disputed the claims that metaphysics is impossible and that science does not raise metaphysical questions. A third argument involves defending the necessity of introducing trans-physical entities or qualities in order best to explain the physical world we find around us. One way to do this is to make the case for emergent properties (see Chapter 8 below). Another way is to argue from the sense of transcendence which, many argue, is basic to the human response to the world. If the human spiritual impulse is focused not on the world alone (viewing *it* as spiritual) but moves beyond the world, then it is necessary for us to locate the source of spirituality in a being or realm beyond the physical universe.[58] If this task should entail speaking in theistic terms, as I think it does, then there is no need to resist the sort of theological reflection required to give shape to these intuitions.

SOPHISTICATED (RELIGIOUSLY TINGED) NATURALISM

As an example of sophisticated (religiously tinged) naturalism we can turn to one of the most sophisticated scholars in the discipline of religion/science writing today, Willem Drees. Drees begins with the premise that all elements of reality are constituted by the same kinds of matter (ontological naturalism). 'Higher' or more complex phenomena, such as religion and morality, are thus to be viewed as natural phenomena; they may have their own concepts and explanations, but should in no way be considered as *super*natural: 'The challenge is to accommodate religious positions not merely to contemporary physics, but also to insights gained through evolutionary biology and the neurosciences' as well as the social sciences.[59] At the same time, Drees insists that he is not eliminating or reducing the importance of religion: 'Religious traditions are phenomena which differ from physical characteristics in that they embody an awareness of a reality which is different from the reality of our daily lives.'[60]

In his major book on physical cosmology and its implications for religion, *Beyond the Big Bang*, Drees provides a more sophisticated account of the possible relations between cosmology and theology than is usually offered:

> Identifications (like Jastrow's [who argues in *God and the Astronomers* that science leads to theology]), contrasts (like Sagan's [who suggests at one point that Hawking's physics leaves no room for a Creator[61]] and Gribbin's [who argues that physics excludes metaphysics[62]]), or other implications of recent cosmology for theology and metaphysics can be found in abundance, especially on the last pages of popularizations of science. Such statements are not as sophisticated as the science that has been drawn upon in the preceding pages. These glib statements express much too simple views of how science and theology interact. They are unsatisfactory for all the fields involved: philosophy, theology, and science. Such statements are often philosophically naive, as they neglect the methodological problems arising in the transfer of ideas from one more or less coherent system of language and thought to another. Theology and metaphysics are misunderstood as pre-scientific answers to questions now better answered by science – and therefore abandoned. This view of theology as an unsuccessful attempt at explanation is, however, not in line with theology's present-day conception of its own task.[63]

Drees and I may be on opposite sides of the religious fence, since the present book is a defence of theology in the face of the naturalism that thinkers like Drees take as given in 'the age of science'. But I must nonetheless underscore his model of the discussion between science and theology. Drees sees that it is not possible to move directly from physical conclusions to theological ones – whether positive *or* negative – for there are 'methodological problems arising in the transfer of ideas from one . . . system of language and thought to another.' This was a central thesis of my *Explanation from Physics to Theology*: the move to theological explanations must include, among other things, the concern with contexts of human meaning basic to the social sciences.

Drees (often) realises that many of the meetings and negotiations between science and theology concern questions which are fundamentally philosophical in nature. What he has not seen quite as clearly is that many of these questions are trans-physical and therefore not decidable based on scientific results alone. So, for example, the question whether *naturalism* is the most adequate framework for discussing a whole range of topics – from physics through the study of the human animal and its functioning and on to the phenomenon of religion – is not one that can be decided by amassing data. It is 'metaphysical' in the sense that one can resolve it rationally only when one carefully formulates on a meta-level the whole range of interpretations of the physical data. To begin with the appeal to ontological naturalism is to rule out viable alternatives from the start – alternatives such as emergentism, theism, panentheism and pantheism.

Nonetheless, Drees's naturalism does represent in some respects the theologian's most serious competitor. This view dispenses with miracles, and indeed with beliefs about God and the supernatural altogether; yet it claims to reserve an adequate place for religion and for awe as the religious response to the universe. Such a view is more parsimonious than even the minimalist religious positions we have looked at in the preceding pages (e.g. Tilby, Harrison), it presents no conflict with science whatsoever, and yet it claims not to be dismissive toward the phenomenon of religion. Drees's, for instance, is fundamentally a functionalist account of religion, which he views in terms of human needs and the human quest for meaning. Unfortunately, his naturalism precludes any serious interest in the *content* of theological assertions; much more important to him is the quasi-Darwinian question of what 'survival value' religion has for humanity. Like an evolving species, religious beliefs alter over time as a result of environmental changes and humanity's development.

So far in this chapter we have encountered three competing claims for primacy: the primacy of spirituality and religious experience, Drees's primacy of the evolutionary/functionalist perspective, and my own defence of theological reflection on the content of religious beliefs. My concern with the functionalist position is that its advocates have difficulty remaining consistently at that level. Drees certainly begins in functionalist terms:

> The position defended here will be that knowledge is a product, a construction made by humans – with their conceptual apparatus, in their mathematical and natural languages – in their encounter with reality. Hence, any consonance is also a construction and not a discovery of a pre-established harmony found in reality. One can construct notions of God which are consonant with the different

cosmologies. The meanings of theological notions become related to a scientific view of the Universe. A more explicit theological conclusion will also be drawn. The expectation of consonance as a feature present in reality was based upon the assumption of God's presence in this world, 'God showing through'. If consonance is not found one might question that presence.[64]

Already here, however, one finds irreducibly theological statements ('God's presence in this world'). The fact that one is to *assess* theological claims, as in his last sentence, by an empirical *modus tollens* – If God exists, theistic language will be 'consonant' with physical cosmology; hence lack of consonance implies God does not exist – does not take away from the *content* of such claims. Moreover, the whole process is ruled by a metaphysical notion, the 'encounter with reality'. Drees seems almost to admit that the functionalist or constructivist approach is controlled by a meta-naturalist notion: 'A mathematical theory which would explain how one could get a universe from nothing would not give a physical universe, but the idea of a physical universe. All evidence is *post factum*. Reality is assumed rather than explained. This applies also to my naturalist account.'[65]

If the controlling principles are metaphysical, how far can we go with them? Drees recognises at one point that one needs to find a level playing field that is not either purely scientific nor purely theological:

Assuming consonance between a theological idea and a scientific theory implies that we search for a suitable interpretation of the concepts involved. *This probably implies that we place both in a (not necessarily complete) metaphysical perspective.* By doing so the terms at all three levels – theology, metaphysics, and science – might change meaning.[66]

Furthermore, Drees sees that one also needs to 'provid[e] grounds as far as possible' for one's views at this (metaphysical) level, which means that one 'assumes the possibility of argument across the boundaries of specific moral and religious views, and hence a metalevel of honest intellectual discourse.'[67] What results is what he calls a 'constructive' model of theological reflection.[68]

Drees is also exactly right that such an endeavour might also supply criteria for scientific decision-making:

What exactly gives way and what remains unchanged depends on the relative importance of the elements and their reliability. In general, scientific knowledge is considered to be the most reliable

153

[= epistemological naturalism?]. However, sometimes science itself allows for different interpretations – as in contemporary quantum theories. And science is changing. At the frontier, where science is dealing with unsolved problems, there is a variety of avenues to pursue – as we will see for the case of quantum cosmology. *In such cases a metaphysics informed by a religious perspective might determine the criteria for theory development and appraisal.*[69]

But after all the talk of a constitutive function for metaphysical reflection, Drees still harbours an inexplicable scepticism about theology:

> My approach in this work is 'from below'. There is no specific source of information about God. Theological concepts are . . . explicitly introduced as assumptions. These notions do come out of a tradition, but in this study their origin – in scripture, revelation, or whatever – is not used as an argument.[70]

One cannot help but worry that it is a straw man with whom Drees is wrestling here. The positions that we have found viable in this and the foregoing chapters do not appeal to unimpeachable foundations for theological knowledge, and most emphatically not to Christian scriptures as inerrant sources of truth. Instead, the postfoundationalist approach to theology that we discovered in Wenzel van Huyssteen and others uses the *fit* and *explanatory power* of theological ideas as important factors in their defence. In fact, Drees himself expresses something very much like the required testing procedure. He writes:

> If ideas about God can be successfully embedded in a network of concepts, the ideas about God receive some credibility from the overall credibility of the network. If we can incorporate the best of contemporary cosmology, and thus much of physics and astronomy [within a theological framework], the whole network then deserves to be taken seriously – if the network is also internally coherent. This justification of credibility by coherence is also present in theoretical science, and certainly in cosmology. It does not justify a strong claim on truth as correspondence between concepts and reality, at least not for those elements at a distance from the most empirical elements, but it suggests that through those concepts we might be in touch with something 'out there'.[71]

If the position argued in this section is correct, we have achieved an important breakthrough. Drees's work shows, like few others, what the costs are of attempting to do theology 'in light of contemporary science'.

The preference for naturalistic accounts, where available, for explaining occurrences in the world – and the concomitant reticence to accept supernaturalistic explanations – become a guiding framework for the doing of theology. At the same time, the scientific conundrums and the limits of science themselves underscore the need for a broader, a *meta*physical discourse. Science, which works so well as an explanatory agent, functions poorly as a world-view that might guide moral decision-making and the human quest for meaning. Religion functions well in this regard – but only if it is analysed not only in terms of its functions but also in light of its actual claims about the divine. Finally, we find that the fit and explanatory power of theological ideas recommend them as viable claims about the nature of reality. That theological theories are also human constructions need not be disputed, as long as this does not rule out the possibility that they may provide the most powerful explanations we possess of the world of our experience taken as a whole.

IMPLICATIONS OF THE DISCUSSION

We have discovered a continuum in the positions on the science/theology dialogue about cosmology.

1. At one extreme we find a sort of natural theology: thinkers like Jastrow argue that science leads to theology.
2. Others hold that science (contemporary physics) supports one particular religious viewpoint. Unfortunately, they disagree as to what it is. The natural theologians argue, for example, that the best answer is some form of theism. Some of their opponents[72] find a closer connection with Eastern metaphysics, others with generic spirituality, and yet others with metaphysical views that are thoroughly non- or even anti-religious.
3. The next group of thinkers maintains that science by itself amounts to a sort of religious perspective. Thus Tipler in *The Physics of Immortality* finds within physics alone sufficient reason to expect a sort of immortality for intelligent agents, without (he thinks) needing to move beyond physical theory to a religious perspective.
4. Others hold that science supports *multiple* religious perspectives. Paul Davies, for example, finds arrows in the direction of a religious interpretation of the universe, though he would insist that these indications underdetermine the choice between particular religious traditions.

5. More (metaphysically) sceptical are those who find spirituality in or implied by science and its results while resisting making any truth claims on this basis (Harrison, Tilby).
6. More sceptical still are the advocates of science and theology as two distinct activities which have nothing to do with each other.
7. Finally, others hold that theology is a pure construct, and naturalism represents the best truth we have about the world.[73]

An important insight that has emerged out of the survey of alternatives in this chapter (particularly in the treatment of Willem Drees[74]) involves the use of some form of metaphysical discourse for mediating the debate. Again and again, we saw the need for a 'third party' to provide a neutral context within which claims could be evaluated. How else are we to adjudicate such radically discrepant claims as, on the one hand, the claim that science gives evidence of God's existence and, on the other, the claim that science has abolished once and for all superstitious doctrines such as theism?[75] What one needs is *not* a new discipline that will pronounce with authority on the questions. Instead, what is required is a common framework for formulating agreements and disagreements – one within which common terms and definitions can be found for presenting the whole spectrum of views. Only then can their divergences (and the best arguments for and against each one) be clearly recognised.

In pursuing 'descriptive metaphysics' in this sense,[76] one does not need to expect resolution of all the differences through argument alone. Theologically, the opening four chapters have made it clear enough why convergence through 'universal reason' should not be expected. Scientifically, there is an additional reason, however: metaphysics represents, at its best, something like long-term science – speculation about how it will all fit together in the end. Since such reflection goes beyond what we can *currently* nail down through agreed-upon evidence, there will obviously be unresolved differences. Nevertheless, our *in*ability to compel others to our particular versions of how it will all turn out in the end in no way justifies abandoning the project altogether. To do so would be harmful both to science and to theology.

A THEOLOGICAL RESPONSE

We have surveyed physical cosmology and the major religious responses to it and have come away without a clear apologetic victory for Christian theology. Attempting to move beyond superficial appropriations of the data, we have instead found a rich array of theories, with hotly contested debates currently raging over the best overall framework for making sense of them. In the end, we were unable even to move from the most recent scientific picture of the world to theism (in some form or another) as the only viable interpretive framework.

Still, the discussion has not proved wholly without positive results. Big Bang cosmology is clearly closer to the biblical position of an initial creation by God than is Steady State cosmology. Further, we have found scientists insisting time and time again that the conditions and constants lie within that one unbelievably narrow range which alone would allow for the eventual emergence of life in the universe. Even those with no theistic leanings and no love for metaphysics speak of the *appearance* of teleology or goal-directedness in the universe – although they are quick to add that it is 'as if' the initial conditions were set up for a purpose.[77] Whatever other services this chapter may provide for theologians, it should at least help to guard against the over-quick movement from scientific results to conclusions about the nature of God and his relation to the world. We have not been able to find clear errors, for example, in pantheistic or spirituality-based responses to recent cosmology. Eastern metaphysics is not obviously inferior to theistic metaphysics when it comes to accounting for the data. And Willem Drees serves as a constant reminder that there is much to be said for a naturalist metaphysics that views all trans-physical reflection with caution and seeks to delineate it sharply from science as 'constructs' from 'hard facts'.

What does this all say to the theologian who begins with belief in God and turns to what humanity knows about the natural world in her quest for understanding? First and foremost, she does not find confrontation or falsification. Thanks to scientific developments in this century and the supersession of positivist and foundationalist theories of knowledge, the theologian may begin unapologetically with her faith in theistic creation as recounted in the scriptural stories. Indeed, we have even found *some* reason to think that belief in the threefold Creator God is better off than it has been any time since Newton's *Principia mathematica* was published in 1686. In the non-physical spacetime before $t = 0$ or after the Big Crunch, ample (theological) room remains for the One in whom there is no shadow or turning of days. It will be our task in later chapters to find place for the divine providential care of the world as well.

What I have tried to convey is at least an initial sense of what science does and does not provide. Only when these parameters are in place can a clear picture be given of what it is that theology adds to the natural scientific knowledge of the universe. The net result is an interweaving of scientific and theological components into a rich account of what is involved in the creation and sustenance of the world by God. I summarize this theology in eight theses:

1. God's time is not our time because, assuming God pre-existed the world (a belief I take to be required by any version of theism), he *could not* have existed in anything like the (physical) time that characterises the physical universe. Conceiving God as an agent requires that this pre-universe existence not be understood as a state of timeless eternity in the sense of Augustine. Yet clearly the passage of time in the divine life must be marked in a manner very different from the one we know. Christian theology will naturally seek to comprehend divine time in terms of the inner dynamics of the divine triune being. Presumably, time in the eschatological state after the end (or transcending) of the universe must be understood in a similar fashion.

2. God created the universe in a singularity of matter/energy, a point infinitely small and infinitely dense. Physical space and time were not created separately or prior to the 'moment', as Newton thought, but rather emerged along with the creation of the world itself. The universe does not expand into empty (physical) space, but space itself expands as the universe unfolds. Nonetheless, the universe begins and ends within the divine being, outside of which there is truly nothing.

3. The universe evidenced lawlike behaviour (mathematical regularities) from the very beginning, although the laws that controlled the expansion in the very first instants were very unlike the laws we now observe in the universe.[78] *That* the fledgling universe was lawlike and exhibited regularities of motion is, theologically speaking, an expression of the regular and rational nature of God. Even *what* those laws were can be correlated with, though not deduced from, the nature of God.[79] One expression of the lawlike nature of our universe is the physical constants that express the regularity of physical interactions. Viewed theologically, these too become expressions of the constancy of the character of God.

4. From the beginning, this universe was such that it was possible, if not inevitable (this, I think, we do not know), that life would someday emerge. Further, it was possible though not inevitable (this also, I think, is unknown), once life emerged on our planet, that it would someday evolve into intelligent life forms – the sort that could be reading and comprehending a sentence like the

present one. Since we believe theologically that God created the universe as a habitation for life and ultimately for humanity, we believe that the existence of these conditions was from the start an expression of divine providence. Speaking from the standpoint of faith, we must say that God created the universe with the eventual emergence of life in mind.

5. The net result of these regularities (3–4 above) is that the universe appears to the physical scientist – at least to the one who takes *her own existence as inquirer* as one of the data to be explained – as though it were designed *in order that* humanity (and perhaps other intelligent life forms) could one day emerge. I expressed scepticism about arguments 'from below' (so-called anthropic arguments) that attempted to prove the inevitability or necessity of intelligent life someday arising in this universe. Such scepticism does not apply, however, to theological observations that are 'from above' – or, to put it in less imperialistic sounding terms, to those that are *ex fidei*, from faith. Theologically, one can only conclude that the appearance of order in the universe is precisely what we would expect in a universe designed by God for life to emerge and prosper.

6. In theological terms, one will wish to say much more about the God/world relation than can be said on the basis of the evidence deduced so far. Above all, the theologian will wish to affirm, 'Behold, it was (is) good,' because the nature of the One who created it *is* to be good, and indeed to set the standards for goodness. The Christian proclamation is that everything that has occurred – including the very costly 'means' of evolution, which involved incredible suffering – somehow works together for good (Rom. 8: 28). This statement may not be made in a Polyanna-esque fashion. In the face of suffering it is a gut-wrenching claim, better made through tear-filled eyes than in joyful exuberance. It means that, despite appearances to the contrary – and sometimes, it seems, against all evidence – theology asserts that the final result, even when taken together with the means by which it was achieved, will be good. Such a judgement can never be 'justified' by scientific means, although it does presuppose a knowledge of cosmic history that science alone can supply.

7. Theology also makes assertions about the place of humanity in the cosmos that cannot be derived from science. Specifically, it asserts that humanity has a moral nature, that it is made specifically in the image of God, and that God has special purposes with humanity. But it is not surprising that the natural sciences would offer little

159

help here, since the claims in this category fall in the domain of the human sciences rather than the physical sciences.

8. Finally, theology makes certain claims about the future of the universe that cannot be derived from anything we currently know through science (though, again, eschatology presupposes some knowledge of the physical cosmos and its probable future). In particular, Christian theology turns on the promise that God will someday complete the work that he has begun, drawing all things into the divine presence and perfecting the work he began at creation. For the Christian this occurs 'according to [God's] purpose which he set forth in Christ as a plan for the fulness of time, to unite all things in him, things in heaven and things on earth' (Eph. 1: 9f.), in so far as Christ 'is before all things, and in him all things hold together' (Col. 1: 17). This completion includes the redemption of the world, which 'has been groaning in travail together until now' (Rom. 8: 22), as well as of the persons within it. Ultimately, there will be 'a new heaven and a new earth, for the first heaven and the first earth [will have] passed away' (Rev. 21: 1) – a Sabbath rest in the presence of God (Heb. 4) in which a direct relationship with God will be possible.

CONCLUSION

It is possible to feel some disappointment that a study of the physical universe does not yield more evidence in defence of these fundamental theological theses. Three things may be said in response. First, science *has* been able to provide some constraints on theological theory formation, and hence on a Christian theology of creation. Second, science cannot speak to human agents about their quest for meaning in the world unless and until its results are placed within a broader (metaphysical) context for discussion. Third, *when* we enter into that sort of debate, I believe that we find that theism in general, and panentheism in particular, is able – and I believe is *best* able – to integrate the scientific results with what we know of our existence as human beings in this world. The starting point in the explanatory project may be scientific, but the whole picture involves incorporating the religious dimension as well.

Perhaps surprisingly, then, our conclusion has been that *it is not the particular conclusions in physical cosmology that are most helpful to theologians.* In fact some results of our study were helpful while others were either neutral or negative for the theological enterprise. Instead, *the single greatest positive result of current discussions in cosmology lies in the fact that scientific results plead for meta-physical, and ultimately theological, treatment and interpretation.* Let me put the point differently: theologians may need to be good *listeners* when it comes to new discoveries about the physical world, since these provide knowledge of that universe whose existence we attribute to God the Creator. Knowledge of the universe forms the backbone of (the empirical side of) the doctrine of creation; it gives us a fuller sense of 'the world which Thou hast made'. By contrast, theologians need to be key players when it comes to the *interpretation* of those scientific results. For the process of interpretation, though constrained by science, is not dictated by science. We have begun to see that the theologian has much to say about the broader interpretive questions – and especially about the nature of the being that turns its telescopes and microscopes upon the universe around it.

NOTES

1. Again, as I wrote in the Preface, I recommend study of the Vatican Observatory volumes edited by Robert J. Russell and of monographs such as Arthur Peacocke's *Theology for an Age of Science* (Minneapolis: Fortress Press, 1993) for a more detailed presentation of the concrete data. Excellent presentations of the scientific theories can be found in the various books of John Leslie and John Barrow, and slightly more popular accounts in Paul Davies' publications.

2. I have been tempted to add under each heading long lists of thinkers whom the heading might subsume. Though such lists can be helpful, they also tend to result in superficial treatments of sophisticated and subtle thinkers – and, worse, to superficial dismissals of careful arguments. Better for the reader to note the major categories or 'types' in what follows and, after careful study of a particular text, to make her own assessment of the categories under which it might be fit.

3. See Robert Wesson, *Cosmos and Metacosmos* (La Salle, IL: Open Court, 1989).

4. Wesson, *Cosmos and Metacosmos*, pp. 27-52.

5. See, e.g., Wesson, *Cosmos and Metacosmos*, p. 11.

6. Angela Tilby, *Science and the Soul* (Melksham, Wiltshire: Cromwell Press, 1992), p. 2.

7. Tilby, *Science and the Soul*, p. 193.

8. See Tilby, *Science and the Soul*, Chapter 10.

9. The use of metaphor in theology does not *need* to exclude the truth question. See Janet Martin Soskice, *Metaphor and Religious Language* (New York: Oxford University Press, 1985), and Ian Barbour, *Myths, Models and Paradigms* (London: SCM Press, 1974). On Paul Ricoeur's famous doctrine of 'metaphorical truth', see Kevin J. Vanhoozer, *Biblical Narrative in the*

Philosophy of Paul Ricoeur: A Study in Hermeneutics and Theology (Cambridge: Cambridge University Press, 1990).

10. Tilby, *Science and the Soul*, Chapter 10.

11. Tilby, *Science and the Soul*, pp. 224–6, 229f.

12. Tilby, *Science and the Soul*, Chapter 8.

13. See Michael Redhead, *From Physics to Metaphysics* (Cambridge: Cambridge Univ. Press, 1995).

14. Tilby, *Science and the Soul*, Chapter 9.

15. Cf. John Barrow and Frank Tipler, *The Anthropic Cosmological Principle* (New York: Oxford University Press, 1986).

16. Arthur Peacocke, *Theology for a Scientific Age: Being and Becoming – Natural, Divine, and Human*, enlarged edition (Minneapolis: Fortress Press, 1993), p. 72.

17. Frank Tipler, *The Physics of Immortality: Modern Cosmology, God, and the Resurrection of the Dead* (New York: Doubleday, 1994).

18. Frank Tipler, 'The Omega point theory: A model of an evolving God', in Robert J. Russell, William R. Stoeger, SJ, and George V. Coyne, SJ (eds), *Physics, Philosophy, and Theology: A Common Quest for Understanding* (Vatican City State: Vatican Observatory, 1988), pp. 313–31.

19. To name just one example: John Gribbin's book, *In the Beginning* (Boston: Little, Brown, 1993), starting from the same set of scientific data, defends, among others, the following conclusions. Life is not a concept that can be isolated; it is defined by interrelation, and reproduction is its most important trait (pp. 46–8). Earth was 'seeded' with organic molecules (or perhaps self-replicating molecules) from outer space (pp. 69ff.). The so-called Gaia hypothesis (e.g. p. 112) holds at the galactic level: galaxies are life forms of some sort, as is attested by their spiral form (which also proves the existence of 'dark matter' (pp. 200, 206, 213). Stars are a sort of foam on the ocean, affected by dark matter. Cosmologists observe a creative process, watching stars being born and forming galaxies. Universes generally come into existence for only a few seconds because of quantum fluctuations. But those universes that include black holes reproduce; by Darwinian principles, these will be the ones that continue to exist and reproduce (pp. 251–3); hence universal evolution selects for those universes that have black holes. By coincidence, these are precisely the ones (the only ones) in which life can evolve. Our universe is closed (pp. 231f.) and is inside of a black hole (p. 242). (Actually, within every black hole is a universe.) Again, universes are 'alive' (p. 252) and they reproduce; those same forces that lead to self-replicating universes also happen to suffice for the production of life forms such as us. At one point Gribbin quotes James Lovelock: 'Life was thus an almost utterly improbable event with almost infinite opportunities of happening' (p. 255).

20. It is helpful to contrast Tipler's approach to the work of C. J. Isham, who insists that one can't derive religious implications directly from science (see Isham, 'Creation of the universe as a quantum process', in Russell, Stoeger and Coyne (eds), *Physics, Philosophy, and Theology*, pp. 375–406.

Isham explores the implications of a quantum-based creation theory as it applies to different mathematical formulations of space-time, time, etc. His major emphasis lies on the 'Hartle/Hawking' hypothesis, which considers time a fourth dimension that only comes into 'focus' (in a Newtonian sense) 'later'. There is no 'point of creation' on the Hartle/Hawking model. No one can work through Isham's article without realising how difficult it is to ground particular theological conclusions directly on the scientific evidence. The mathematical theories, even apart from their metaphysical implications, are complex enough that it is clear that they will have to be mediated through philosophical assumptions before they can become theologically fruitful — or even threatening to theology!

21. Paul Davies, *God and the New Physics* (New York: Simon & Schuster, 1983).

22. See Fritjof Capra, *The Tao of Physics: An Exploration of the Parallels Between Modern Physics and Eastern Mysticism*, 3rd edn (Boston: Shambhala Publications, 1991), and Gary Zukav, *The Dancing Wu Li Masters: An Overview of the New Physics* (New York: Bantam Books, 1980).

23. For a number of years Wolfhart Pannenberg has defended the parallels between the Judeo-Christian notion of spirit and fields of force in modern physics. See his doctrine of creation in Pannenberg, *Systematic Theology*, trans. Geoffrey W. Bromiley, 3 vols (Grand Rapids, MI: Eerdmans, 1991), Vol. 2, Chapter 7.

24. Paul Davies, *Superforce* (New York: Simon & Schuster, 1984), p. 268.

25. Paul Davies, *The Cosmic Blueprint* (New York: Simon & Schuster, 1989).

26. Arthur Peacocke has also argued in various publications, *pace* Monod, that the random variation and selective retention of biological evolution is an ideal means for increasing complexity; see, for example, *Theology for a Scientific Age*, Chapter 4. Peacocke does not accept the conflict between biological and physical (thermodynamic) laws, although he does employ the framework of emergence in speaking of biology.

27. Paul Davies, *The Mind of God* (New York: Simon & Schuster, 1992), p. 232.

28. Cf. the distinction between 'prediction' of future findings by a theory and 'accommodation' of existing data in Peter Lipton, *Inference to the Best Explanation* (London: Routledge, 1991).

29. Robert Jastrow, *God and the Astronomers* (New York: W. W. Norton, 1992).

30. Jastrow, *God and the Astronomers*, p. 21.

31. Peacocke, *Theology for a Scientific Age*, p. 134.

32. See, for example, Ted Peters, *Fear, Faith, and the Future: Affirming Christian Hope in the Face of Doomsday Prophesies* (Minneapolis: Augsburg, 1980); Peters, *God – The World's Future: Systematic Theology for a Postmodern Era* (Minneapolis: Fortress Press, 1992); Peters, *Genes, Theology, and Social Ethics: Are We Playing God?* (Berkeley, CA: Center for Theology and the Natural Sciences, 1994).

33. For a full account of his reflections on the genome project and its implications, see Ted Peters, *Playing God? Genetic Determinism and Human Freedom* (New York: Routledge, 1996).

34. Ted Peters (ed.), *Cosmos as Creation: Theology and Science in Consonance* (Nashville, TN: Abingdon Press, 1989).

35. Ted Peters, 'Theology and natural science', in David Ford (ed.), *The Modern Theologians: An Introduction to Christian Theology in the 20th Century*, 2nd edn (Oxford: Blackwell, 1997).

36. Ted Peters, 'On creating the cosmos', in Russell, Stoeger, and Coyne (eds), *Physics, Philosophy, and Theology*, pp. 273–96.

37. Stanley L. Jakti, *Cosmos and Creator* (Edinburgh: Scottish Academic Press, 1980).

38. Jakti, *Cosmos and Creator*, p. 129.

39. Jakti, *Cosmos and Creator*, p. 141, emphasis added.

40. For example, Sir Arthur S. Eddington, *The Nature of the Physical World* (New York: Macmillan, 1928); Werner Heisenberg, *Physics and Philosophy: The Revolution in Modern Science* (New York: Harper, 1958); Carl F. von Weizsäcker, *The World View of Physics*, trans. Marjorie Grene (Chicago: University of Chicago Press, 1952); von Weizsäcker, *The Rise of Modern Physics*, trans. Arnold J. Pomerans (New York: Braziller, 1957).

41. Indeed, so much so that people sometimes joke that, just as the Hindu was traditionally supposed to withdraw from his or her responsibilities in the world at 50 and devote himself or herself to meditation, so also the theoretical physicist is to withdraw from active theoretical work in her 50s and write a book on the philosophy (or theology) of physics. See, among the best, Freeman Dyson, *Infinite in All Directions* (New York: Harper & Row, 1988), and Bernard d'Espagnat, *Veiled Reality: An Analysis of Present-Day Quantum Mechanical Concepts* (Reading, MA: Addison-Wesley, 1995).

42. David Bohm, *Causality and Chance in Modern Physics* (Philadelphia: University of Pennsylvania Press, 1957), p. 141. More recently, see Bohm and B. J. Hiley, *The Undivided Universe: An Ontological Interpretation of Quantum Theory* (London: Routledge, 1993).

43. Bohm, *Causality and Chance*, p. 146. A similar approach has been taken by the leading French physicist Bernard d'Espagnat in his recently translated book, *Veiled Reality*. D'Espagnat makes crucial and effective use of the 'holism' of Spinoza.

44. Bohm, *Causality and Chance*, p. 170, emphasis added.

45. David Bohm, *Wholeness and Implicate Order* (London: Routledge & Kegan Paul, 1980).

46. Bohm, *Wholeness and Implicate Order*, p. 49.

47. See Kevin J. Sharpe, *David Bohm's World* (Lewisburg, WV: Bucknell University Press, 1993), p. 57.

48. Sharpe, *David Bohm's World*, p. 61.

49. This is clearest at the end of Sharpe's book; see *David Bohm's World*, especially pp. 96ff.

50. Bernard d'Espagnat, *Veiled Reality*, cited above.

51. See Peter L. Berger, *The Sacred Canopy: Elements of a Sociological Theory of Religion* (Garden City, NY: Doubleday, 1967).

52. Edward Harrison, *Masks of the Universe* (New York: Macmillan, 1985), p. 273.

53. Harrison, *Masks of the Universe*, p. 270.

54. Harrison, *Masks of the Universe*, p. 270.

55. Harrison, *Masks of the Universe*, pp. 216–18.

56. Harrison, *Masks of the Universe*, p. 271.

57. See John Hick, *An Interpretation of Religion* (New Haven, CT: Yale University Press, 1989). Indeed, Hick claims that his religious view transcends even the distinction between theism and pantheism.

58. For an example of the richness of this argument see Louis Dupré's now classic work, *The Other Dimension: A Search for the Meaning of Religious Attitudes* (Garden City, NY: Doubleday, 1972).

59. Willem Drees, *Religion, Science and Naturalism* (Cambridge: Cambridge University Press, 1996), p. xii.

60. Drees, *Religion, Science and Naturalism*, p. 281.

61. See Carl Sagan's 'Introduction' to Stephen Hawking's *A Brief History of Time* (Toronto: Bantam Books, 1988). Against Sagan's 'atheist' interpretation of Hawking's physics see David Wilkinson, *God, the Big Bang and Stephen Hawking* (Turnbridge Wells, Kent: Monarch Publications, 1993). Wilkinson looks in chapters 1–8 at the scientific understanding of the universe and in Chapter 9 at the biblical understanding of the universe (see also the Appendix). In Chapter 10 he then argues that Hawking's physics, correctly understood, no more counts against theistic belief than did the physics of Galileo – whatever the church of the time might have thought.

62. See John Gribbin, *In Search of the Big Bang: Quantum Physics and Cosmology* (Toronto: Bantam Books, 1986).

63. Willem Drees, *Beyond the Big Bang* (La Salle, IL: Open Court, 1990), p. 2.

64. Drees, *Beyond the Big Bang*, p. 6.

65. Drees, *Religion, Science and Naturalism*, p. 268. Thus Drees is sceptical of purely quantum-based cosmologies: 'In my opinion, the appeal to quantum fluctuations does not offer a final answer. Any such scheme rests upon certain assumptions. In the case of the idea of origination by quantum fluctuations there is the assumption that some form of quantum laws holds for some form of reality, such as the "pre-geometry" in [Peter] Atkins' account . . . Even the most extreme "nothing" of a physicist is not an absolute Nothing devoid of any properties and measures. Other proposals do not fare better. For example, Stephen Hawking's proposal for a universe without boundaries avoids questions about a beginning in time, but it does not do so without making any assumptions' (pp. 267f.)

66. Drees, *Beyond the Big Bang*, p. 7, emphasis added.

67. Drees, *Beyond the Big Bang*, p. 3.

68. Drees writes, 'If evil is really evil and injustice in this world is not illusory, any proposal for meaning is just that: a proposal, an orientation intended to guide action towards a better world, "dreaming of peace". The constructive nature of theology is not only a limitation of our knowledge; it reflects the nature of the theological enterprise' (*Beyond the Big Bang*, p. 7).

69. Drees, *Beyond the Big Bang*, pp. 6f., emphasis added. He notes the similarities of his theory of 'constructive consonance' with the 'interanimation
theory of metaphors' (Janet M. Soskice, *Metaphor and Religious Language*
(Oxford: Clarendon, 1985)) and with the view of scientific and religious
research as a 'metaphoric process' (M. Gerhart and A. Russell, *Metaphoric
Process: The Creation of Scientific and Religious Understanding* (Fort Worth,
TX: Christian University Press, 1984)). Drees adds, 'Interesting metaphors
are not merely a substitution of scientific terms for theological ones nor
a direct comparison. It [*sic*] changes the interpretations of the concepts
used. It implies that one constructs a new understanding of reality'
(*Beyond the Big Bang*, pp. 6f).

70. Drees, *Beyond the Big Bang*, p. 8. Also in Chapter 6 he argues that there
are no cognitive arguments for the hypothesis of God. He endorses at one
point Nicholas of Cusa's doctrine of 'learned ignorance' (p. 195), citing
the work of Harrison that we explored above. Yet Drees also holds that
acceptance of the God hypothesis is like commitment to a paradigm or
research programme (p. 196), which may well involve actual belief.

71. Drees, *Beyond the Big Bang*, p. 9.

72. For example, Capra, *The Tao of Physics*; David Bohm, whose affinity is with
the holism of Eastern metaphysics we explore above; and Zukav's *The
Dancing Wu Li Masters*.

73. For example, Drees, Harrison. Recall that Drees offered a metaphysics of
naturalism, in which science provides the *cantus firmus* (he is a realist about
science) while metaphysics remains metaphorical and theology a human
construct. See Part II of Drees, *Beyond the Big Bang*, 'Constructing theology in a scientific culture'.

74. See Drees, *Beyond the Big Bang*, pp. 5ff., 175–9 on 'constructive consonance'. Drees treats theology, science, metaphysics and their relationship
in separate sections.

75. See Carl Sagan's last book, *Demon-Haunted World* (New York: Ballantine,
1997).

76. The term comes from Peter Strawson, who used it in contradistinction
to 'normative metaphysics' because of his scepticism about resolving
metaphysical debates through argument. See the 'Introduction' to Peter
Strawson, *Individuals: An Essay in Descriptive Metaphysics* (Bristol: J. W.
Arrowsmith, 1959, 1977).

77. See the more than thirty interviews with scientists reprinted in Henry
Margenau (ed.), *Cosmos, Bios, Theos* (La Salle, IL: Open Court, 1992),
where the impression of a goal-directed universe receives extensive discussion and support.

78. See, classically, Steven Weinberg, *The First Three Minutes: A Modern View of
the Origin of the Universe* (Toronto: Bantam Books, 1984).

79. This is to break with Kant, who attempted to deduce the fundamental
principles of physics from basic metaphysical principles in *Metaphysical
Foundations of Natural Science*, trans. James W. Ellington (Indianapolis:
Hackett, 1985).

PART III

———⁓⁓⁓ΛΛΛ⌘ΛΛΛ⁓⁓⁓———

Towards a Theology of Divine Action

6

THE PRESUMPTION OF
NATURALISM*

The premise of this book is that the question of divine agency has become one of the most difficult and urgent questions facing theologians today. The question is twofold: what resources can be found within theology – within the doctrines of God, nature, creation and redemption, for example – for addressing these well-known challenges? And what changes may be required in the traditional formulations of these doctrines in order to respond to those challenges that we find ourselves unable to solve?

I should note that there remains one way to minimise or avoid the challenge of science to theology: one can define an 'act of God' in such a way that it does not conflict in any way with the findings or orientation of science. Efforts of this sort have been a major component of modern theology, running from the early 'life of Jesus' (*Leben Jesu*) discussions (Lessing and Schleiermacher and their followers) through the various 'quests for the historical Jesus' of our century. Similar efforts have also motivated many of the rethinkings of the doctrine of God in this and the last century. A recent example might be helpful. A group of leading

---

* More than any other chapter of this book, the following argument owes an enormous amount to ongoing discussions with Steven Knapp. Although I have authored this text, some of the formulations (the reader may assume: the best ones) are drawn from our e-mail correspondence, and many other aspects of the argument are the direct or indirect product of our conversations.

theologians addressed the question of divine action in a now classic collection entitled *God's Activity in the World: The Contemporary Problem.*[1] The general strategy taken by most authors was to reconceive the idea of a divine act or 'the God who acts' in such a way that no conflict, because little or no overlap, with science remains. Schubert Ogden gives particularly clear expression to the strategy that many pursue:

> Instead of asking whether theological statements *can* make sense, [later linguistic philosophers] have attempted to determine what sense such statements in fact *do* make by analyzing their function in actual religious and theological speech. If this shift in approach is, as I believe, fortunate, it presumably has a bearing on the procedure of the theologian. It suggests he will do better not to pronounce on the question *whether* 'God acts in history' makes sense, except insofar as an attempt to show *what* sense it makes implies an answer to that question. If his attempt to explain the sense of the statement is successful, then, of course, one can make sense of it; for *ab esse ad posse valet consequentia.* If, on the other hand, his attempt should fail, the possibility will still exist for someone else to show that, and how, the statement makes sense.[2]

The major contributions in the book – including those by Langdon Gilkey, Rudolf Bultmann, Ogden, John Cobb, David Griffin, Gordon Kaufman, Maurice Wiles, Austin Farrer and Etienne Gilson – either find a way to rethink 'acts of God' that, they think, is both religiously cogent and compatible with contemporary science *or*, if they fail to do so, recommend ceasing to speak of acts of God in the world.

The revisionist strategy is not to be taken lightly: some rethinking of theological doctrines *is* required, and I draw on some contemporary theologians' redefinitions in the following chapters. The present task, however, is a slightly different one, namely: What happens if one begins with divine actions in something like the traditional sense of the word? How strong should the presumption against them be for men and women living in today's world? Are they *ruled out* by science? Are they *compatible* with scientific conclusions? Could there ever be scientific (e.g. historical) evidence that some such acts have actually occurred? What, if anything, causes the conflict between science and divine-action claims – is it scientific results, scientific methods or the scientific mind-set itself? Or could it be that science as such has nothing whatsoever to say for or against talk of divine action?

170

THE PRESUMPTION OF NATURALISM

Perhaps more than anything else, the discussion between theology and science today is concerned with the presumption of naturalism; where it is not, it perhaps ought to be. By the presumption of naturalism I mean the assumption, for any event in the natural world, that its cause is a natural one as opposed to a supernatural one.[3] If, walking through the forest, I hear a branch break from a tree and observe it falling to the ground, I assume that its cause was a weakening of the wood and that it was propelled to the ground by gravity, not that God ripped it from the tree and tossed it to the ground. And the same applies to at least most, if not all, of the other events that I observe or take to have occurred within the natural world.

Indeed, it seems necessary that even the theist should make *some* presumption of naturalism. Not only would it be difficult to imagine any type of scientific activity if one believed it were as likely as not that a given event (say, the branch falling) was caused by God; a full parity between natural and supernatural causes would also make the everyday activities of human beings difficult, if not impossible. When I watch the bedroom door slam shut without being touched by any human being, I naturally assume that the wind did it. Even if I believe that God some- times shuts doors or accomplishes other miraculous events for purposes of his own, this *could not* be the first explanation that I consider when I observe an event in the world. Only when I have ruled out other possible explanations (the wind, a hidden spring, an earthquake) should I turn to the possibility that the event must therefore have God as its cause.

It is surprising to note how often treatments of divine agency over- look the importance of this presumption. A recent collection of essays, for example, is devoted entirely to the question of 'the God who acts'.[4] A number of authors attempt to make a positive case for the metaphys- ical possibility of divine action; the same is true of an earlier (but no less influential) collection of essays on divine action.[5] Yet in all these essays, one scarcely finds a defender who acknowledges that the case *for* divine action hinges as much on overcoming the presumption against divine causes as explanations of events in the world as it does on working out the metaphysics of such action by such an agent.

A similar tendency can be found in a recent work by Nicholas Wolterstorff, who can by no means be accused of naiveté regarding epis- temological issues. In the course of his defence of a woman's 'entitlement' to believe that God is giving her a message for her pastor, Wolterstorff condones her decision to consult a psychologist when she finds herself convinced that God has been speaking directly to her; only after this test has convinced her that she is 'okay' can she act on the basis of the belief that God has given her a particular message for a particular person.[6] Yet whatever presumption motivated this external check plays little (if any) role in the rest of the book, which spells out the mechanisms of 'divine discourse' free from any presumption in favour of natural explanations of (say) the formation of the biblical canon.

The question, I will suggest, is not whether there is any presumption of naturalism but *how strong* we should make it and in what areas we should regard it to be strongest. After a brief look at the way in which the presumption of naturalism has developed in modern (Western) history, I will attempt to state in what senses naturalism is justified and not justified and to assess its significance for the claims of theism about divine action in the world.

HISTORICAL CHANGES IN THE PRESUMPTION

I have suggested that there is a commonsense initial presumption in favour of naturalistic explanations. This argument applies, it seems, to all historical epochs, even those which cannot be accused of enslavement to a secular or scientific mindset. The Israelite prophet Samuel, when he heard a voice in the temple, first went to his father to find out if it was he that was speaking; only when he realised, after multiple trials, that the speaker was not his father, did he become open to the possibility that it was Yahweh (1 Sam. 3). Anthropologists have studied tribal peoples who attribute a number of events to the action of spirits; but even in these cases the first explanation that the tribesperson seeks is usually one of natural causes.[7]

Still, something *has* changed, and that is the explanatory success of natural science. Not until the seventeenth century did we possess a body of theory that was able to explain most physical events in terms of natural laws, laws which, when combined with a clear statement of the initial conditions, allow for predictions of concrete future outcomes. Now it would be a matter of metaphysical prejudice to conclude from this development that it is *impossible* that a given event in the world could have anything other than a natural cause. Nonetheless, it remains a basic feature of the context in which we must do theology today that no other competitor is in the position to provide what the natural scientist can provide: a law-based framework for explaining events ranging from the development of galaxies through the chemical properties of sodium to the causes of brain disease. Again, the emergence of a predictive, nomological science unparalleled in its ability to explain and predict events in the world does not rule out the possibility of divine action in the world; to draw this conclusion would be to take a metaphysical leap that could never be justified on the basis of science's success alone. But such success does establish an *epistemic* presumption in favour of natural explanations – a presumption, that is, on behalf of their superiority for providing us with knowledge of what occurs in the physical world. Since the ability to predict natural events according to absolutely general laws did not exist (say) at the time at which the biblical writers wrote, they did not feel – nor should they have felt – precisely the same presumption of naturalism that we experience today.

Has the presumption in favour of naturalist explanations then continued to increase in a steady climb throughout human history? No, sometimes the task for the defender of divine interventions is easier at a later than at an earlier time. For example, it is (arguably) easier to make the case for divine actions today than at any time since Newton. During the eighteenth century, with the advent of (what seemed to be) a fully predictive science of the physical world, scientists and philosophers concluded that divine action had become all but impossible. But, as we are now discovering, they were doubly wrong. On the one hand, they mistakenly believed that all scientific explanations would be deterministic, that 'completed science' would never countenance non- or underdetermined events. Yet the development of quantum mechanics and its indeterminist picture of the world have falsified this belief; gone is the dream of the Laplacian demon who, given the location and momentum of all particles in the universe at some moment in time, could predict all past and future states. On the other hand, the early modern scientists were mistaken in confusing a methodological presumption with a metaphysical necessity. The success of science does not establish the impossibility, metaphysically speaking, that divine agency might be the cause of some events in the world; with regard to the metaphysical questions, the situation is much the same today as it was several centuries ago. No amount of success in explaining the world through one type of cause can prove that there *could not* be another type of causal force at work in the world.

The presumption of naturalism varies, then, with the nature of our best account of the world – and with the methods that seem to work for obtaining knowledge of the world – at any given time. We might put this point differently: *the most serious presumption of naturalism is methodological.* If an account of natural events in terms of supernatural causes is in conflict with the pursuit of scientific explanations, and assuming (as we have argued) that scientific explanations give us an unparalleled understanding of and control over events in the natural world, then some presumption against supernatural explanations of physical events is unavoidable. I suggest that precisely this is the case. Progress in science is possible, methodologically speaking, only through the prima facie assumption that, for any given event E within the world, E has a natural as opposed to a supernatural cause. Imagine that E is some event whose cause (or the mechanism by means of which it occurs) we don't know – say, prior to the 1950s, the problem of how cells are able to reproduce themselves. Only the assumption that there *must be* some natural explanation can motivate the sort of scientific research that eventually led to the discovery of the DNA structure by Crick and Watson. What holds for the discovery of the double helix holds also for crucial events in the history of evolution: the formation of the first cells within the primal soup, the formation of multiple species from a single common ancestor, and so forth. To maintain that these events might have a supernatural cause rather than natural causes is to stymie science in precisely the region of its greatest potential progress. It is for this reason, and not because of any metaphysical prejudice, that a presumption of naturalism is intrinsic to the pursuit of science as we know it.

THE PARITY BETWEEN PERCEPTUAL AND RELIGIOUS EXPERIENCE

There has been much talk of late of a 'parity' between perceptual experience and religious experience: just as the perception, under standard conditions, of a lake spread out in the valley before me fully justifies me in believing that there is in fact a lake there, so also the perception of divine love in the context of a church service fully justifies me in believing that I am being loved by God, and hence that God exists.[8]

The developments sketched in the previous section show at least one respect in which the parity thesis is mistaken. Even in ancient times men and women accepted a mild presumption of naturalism, in so far as they began with the assumption that individual events in everyday experience had natural causes. Whereas the immediate 'best explanation' for my seeing a tree is that there is a tree in front of me, Samuel could only judge that God was speaking to him after he convinced himself that it was not the voice of his father calling from the next room. Aside from this weak presumption, however, ancient writers may well have been just as justified in accepting a supernatural explanation for some event or series of events as they were in accepting natural explanations. (Indeed, the supernatural framework may well have had *greater* predictive accuracy than the 'naturalist' explanatory principles of their day!)

The situation is different for us. Perceiving that axe strokes caused the tree to crash to the ground immediately justifies the corresponding belief about what actually occurred, at least as long as the implicit explanation is consistent with what we know of nature. But what of the religious 'perception' or sense that God caused the boulder to veer just to the left of the tent in which the children were sleeping? Here the 'perceived' explanation conflicts with the terms in which physical explanations are usually given. If the claim that God altered the rock's trajectory is correct, there is no point in any further scientific research into the incident, because its cause has been made inaccessible to *any possible* scientific reconstruction of the causal history involved. (It is interesting to note that both Plantinga and Alston, in the works cited above, limit themselves to religious perceptions in the realm of the psychological, e.g. one's sense of being loved by God. We return below to the significance of this move.) I conclude that the immediate perception of some state of affairs *in the physical world* justifies an explanatory account of what has occurred *only when* the perceived state of affairs and its alleged cause are compatible with the relevant natural scientific accounts.

We must therefore abandon the parity thesis: whatever may be the limitations on the presumption of naturalism, it *is* strong enough to block a full parity between naturalistic and supernatural explanations for events in the physical world. Humans have been vastly successful at reconstructing the causal histories of events, at least those in the domain of the natural sciences, and at subordinating them to laws that express and explain the regularities in our experience. Given this success, and pending parallel success – or a similar type of universal agreement – in the religious sphere, appeals to divine interventions in the physical world bear a special onus.

WHERE RELIGIOUS CAUSAL ACCOUNTS STAND ON EQUAL FOOTING

When, then, can religious explanations have validity? Must they always play second fiddle? Where might they be on equal ground – or even be presumed superior to non-religious accounts? Hume spoke in a famous passage of the 'before' and the 'after', the realms before the creation of the universe and after the end of universal history,[9] as the areas where it is most natural to speculate about divine action. Wasn't he making a mistake by assuming the principles of experience-based induction would have to apply in these areas as well, and hence that theistic explanations faced an onus here also? In my view it was Kant, not Hume, who raised the bigger challenge: the problem is not that we apply empirical induction to knowledge claims here and find them wanting, but rather that perhaps *no* epistemic criteria can get their teeth into claims to knowledge of the before and the after.

What are these realms? Hume had in mind the realm of the pre-natal and the post-mortem: life after death or the pre-existence of the soul. Nowadays we might speak of the moment just before the Big Bang (or just after the Big Crunch) as a natural location to find the hand of God. The theologian Wolfhart Pannenberg speaks similarly of 'universal history', or history taken as a whole, as the locus of revelation.[10] Surely it would be at the level of the whole of history, if anywhere, that God's intentions, purposes and guiding actions would be visible. What generally characterises religious explanations in distinction to scientific explanations is their tendency to make broad, even universal claims about the direction and significance of history; certainly this is true of religious myths across the world's cultures. To believe in God, in contrast to believing in bacteria or taxes, is to believe something about human experience taken as a whole. If we know that God exists, we know something about the pattern or fate of cosmic history. This is the realm in which religious explanations will have their purchase, if they are to have it anywhere, for here scientific explanations are shackled by lack of empirical data and cannot claim epistemic superiority. (Of course, it is possible that the lack of data will also preclude religious explanations from passing as knowledge.)

The most natural locus of divine action, after God's role as Creator, corresponds to this highest level of analysis, to universal history. The most natural claim for the theist to make is that she sees evidence of God's hand in guiding history as a whole. (We shall return to the question of exactly what kind of causal activity God employs in acting at this level.) The pattern of the broader flow of history should reveal God's gradual working out of his overall purposes. The claim is debatable; indeed, atheists have been eager to dispute it and to argue instead that history as a whole reveals no such purpose or direction.

This, then, is my reconstruction of how to read divine agency claims epistemically: believers form beliefs about the direction of history as a whole, based in part on their beliefs about God's existence and overall purpose; and they break no epistemic obligations in so doing. When they speak of God as being active in some particular event, they extrapolate from this overall sense of divine guidance that God must be active in *this* event because he is active in *all* events and because the nature of this event fits the pattern, either in its situation or its outcome.

Under one interpretation, Austin Farrer's famous notion of 'double agency' offers one way of conceptualising this sort of position.[11] Every physical event has a physical cause, and at the same time God is a contributing cause to every event. If God's role is understood in this metaphysical sense – say, as the force that causes contingent beings to continue existing when they might as well not exist at all – then there is nothing contradictory in speaking of divine causes and at the same time accepting natural scientific explanations of physical events.[12] Under another interpretation of double agency, however, the conflict with natural science returns – the interpretation, namely, that God is one of the efficient causes affecting (and effecting) every event. As Thomas Tracy points out in his recent introduction to *The God Who Acts*,[13] the theory of double agency is inherently ambiguous. If taken in the first sense, it stands in no competition whatsoever with natural scientific explanations; taken in the latter sense, however, it envisions a type of continuous divine intervention in the world that is no weaker than the classical accounts of miracles. It is thus unclear to me how appeals to double agency can help to resolve the tensions raised by claims to divine action.

IMPROBABLE BUT NOT IMPOSSIBLE

What happens then to the theist's belief that God has been involved at *specific* moments in human history? Was Hume right that miracles are unknowable? I shall argue that naturalists are wrong about the impossibility of miracles. The key to the argument lies in the distinction between physical miracles – cases in which God is said to be the direct cause of some event in the physical world – and psychological miracles, where God is said to have introduced a thought or attitude into the mental life of an individual. The answers in the two cases are radically different.

The structure of a physical miracle is from the very outset paradoxical. They presuppose, as we have seen, a divine being who is linked to the meaning and course of history as a whole, while at the same time making highly concrete claims about the causes of specific events. Put differently, the result of a physical miracle is a particular physical state of a particular object in the world, a state that is (in principle) physically measurable and testable; yet the cause of this state is not part of the chain of physical events in the world at all, but somehow breaks into that chain from a realm utterly separate from the physical.

The problem here is that properly theological reasons and accounts do not operate at a level concrete enough for identifying individual miracles. Now Christians, for example, have often held that there are theological criteria for deciding whether a miracle has occurred. For instance, a miracle must be, in its character, consistent with what we know about God; it must further divine purposes as we know them through scriptural revelation; it must lead to unity and spiritual growth within the church rather than to (new) divisions. Most fundamentally, the tradition has held, a miraculous event should be part of God's ongoing process of self-revelation. That is, true miracles do not occur merely for their own sake but always contribute to a fuller understanding of who God is and what his broader purposes are in history.

When one looks back over this list of criteria, it becomes clear that they do not have the sort of preciseness that would enable one to say of a particular physical event that it has a supernatural cause as opposed to a natural cause. Note, for example, that the list of theological criteria does not include the criterion, 'This event *could* not otherwise have occurred, that is without divine assistance, given what we know about the natural world.' And rightly so: to label as a miracle every event that we don't currently understand through scientific means would be to define every gap in our scientific knowledge as a place where God is acting. But this would be to doom claims for divine activity to eventual extinction, in so far as every advance in scientific knowledge would further limit the realm of what God has been taken to do. The fact that such a 'God of the gaps' approach has repeatedly characterised theistic claims about God's activity over the last several centuries in no way mitigates its mistakenness. The widespread scepticism among intellectuals about *any* divine action in the world rests in no small part on the impression, fostered by such a strategy, that the 'space' for divine activity has become smaller and smaller with each new scientific advance.

Scientific improbability – the difficulty that science would face in explaining a particular event – ought therefore to play *no* role in determining which events are miraculous and which events aren't. This leaves us, then, with only the theological criteria listed above. But note that those criteria do not possess the specificity that characterised the (failed) appeal to the scientifically improbable. It therefore appears that theological criteria – and they are rightly the only sort of criteria to which one should appeal in determining which events are miraculous – are not sufficient for identifying any *particular* event as a miracle. It is this result which has led theologians such as Pannenberg to say that for the Christian *all* events ought to be taken as miraculous. The existence of any contingent being or entity is dependent upon the creative activity and continuing will of God without which it would not exist at all. Its existence, therefore, is a miracle at every moment; everything that it does and everything that occurs to it is likewise miraculous. For the Christian then, Pannenberg asserts, all events should be viewed as miracles. (He also argues, following Leibniz, that it would be untheological to understand a miracle as a break in natural law, since this interpretation would cast questions on the perfection of God with regard to his ability to create a universe in the first place that would express his will. The interventionist view likewise misunderstands the nature of natural laws as descriptions of the regularities that we perceive in the world: if God intervenes on a regular basis, then this regularity will itself be a new pattern that can be expressed by refining earlier formulations of natural laws or by introducing new ones.)

What we have found then, surprisingly, is a certain *theological* presumption against the miraculous. History as a whole may be miraculous, as may be the existence of any contingent being whatsoever. Yet the presumption, *even for the theologian*, must be that an individual physical event has physical causes. Even while its ultimate existence and ultimate purpose are explained in terms of God's will, its concrete occurrence – the fact that this physical event rather than another event occurred at this place and time – is presumed to have a physical explanation.

What then of knowledge of physical miracles? We have concluded that the natural sciences meet stringent conditions for knowledge of natural events – conditions for lawlike explanations not met by any other discipline – and that no other explanatory system comes close to meeting those conditions in the natural realm. We can therefore only conclude that there could be no knowledge of a physical miracle if one occurred. Strangely, along with the arguments from natural science, we have been driven to this conclusion by theological arguments. The doctrine of creation suggests that God created an order independent of himself and established patterns within it that make it a predictable and reliable environment for its inhabitants. The very fact that God intervenes only occasionally (if at all) in the world turns out, then, to lead to a presumption in favour of regular or natural causes for any particular inner-worldly occurrence. This presumption, when combined with the success of science as a nomological discipline (a contingent fact which has not always been true in the past and may not be true in the future), makes it impossible to say that I 'know' a miracle has occurred at a given place and time in anything like the same sense in which I say that I know that a standard window when hit with a sledge hammer will shatter.

Can One Know a Miracle Has Occurred?

And yet miracles might occur. If we grant (as we have for the sake of this chapter) that God exists and is directing all of history toward a particular *telos*, then it becomes arbitrary to say that God *could not* be intervening in the physical realm at certain points. Theism with a proscription against any divine action in the world would surely be better labelled deism than theism.

More strongly, it seems to me that a theist might sometimes have good reasons to identify an event as a miracle. Her explanation of what God may have been doing in the event may well not meet the standards for a scientific explanation of that event; it is not a matter of two competing knowledge claims going at it head to head. At the same time, the claim that God was causally involved in a particular event is not just a matter of taste or subjective preference, for the believer may well have good reasons for identifying the event as divinely caused. The combination of her own religious experience, her sense of the pattern of divine activity in the church and the lives of those she knows, the records of divine action in scripture and similar considerations, whether or not they pass scientific muster, allow for a reasoned inference, whatever its epistemic status.[14]

My thesis, then, is that believers may form a reasonable *judgement* that God has (say) contributed to the occurrence of a particular event. To judge is to form a cognitive attitude; the judgement may depend in part on subjective considerations while at the same time involving real content and being based on careful, criticisable reasoning. When we judge that Beethoven's Ninth Symphony could not have had two composers because of the 'spiritual unity' of the piece, we engage in a cognitive act which combines reasons with an inner sense or subjective response to the piece. We might not be able to convince a person not affected by the Ninth Symphony in this way that we are correct in our conclusion; yet it seems illicit to forbid us from making use of our impression in forming a judgement as to its probable composition.

Likewise, it would be illicit to forbid the religious believer to draw on what she has experienced in her relationship with God when she is forming judgements about what God has probably done in history in general or at particular moments in history. Indeed, in making a judgement about a miracle in the past – say, the resurrection of Jesus – the believer may well employ detailed arguments about historical events, texts and motivations. These arguments, however, will not be sufficient on their own, in part because they conflict with what we now know to be the laws of the natural world. Hence, the detailed arguments will need to be supplemented by the believer's judgements about the overall intentions of God in history, God's use of the church as a focal point of his self-revelation, the manner in which God is present to human beings (e.g. as mediated through the Spirit of Jesus), and her own individual religious experience. To say that this complex cognitive process is merely a matter of taste – say, in the sense that one might prefer brown shoes over black – seems woefully inadequate as a description of what occurs. At the same time, the believer's judgement that God has broken natural laws in some particular case, past or present, runs into the difficulties that we have covered above. The judgement is therefore not knowledge, though it is more than taste. In short, the presumption in favour of scientific explanations in the natural realm, and an irreducibly subjective component in the case for any specific miracle, keeps us from using the word 'knowledge' of traditional miracle claims.[15]

PSYCHOLOGICAL MIRACLES

What then of psychological miracles? Can we have knowledge that God is directing the thoughts of an individual, giving her hope or wisdom? Note that the argument as we have developed it to this point does not pertain in the same way to the human sciences. None of the human sciences – psychology, sociology, cultural anthropology and others – has been able to formulate and corroborate fundamental laws of human behaviour. In none of these disciplines is it possible to predict a certain behaviour in the strong sense in which it is possible to predict the behaviour of a physical body in natural science. How, then, should we view the claim an individual might make that God has spoken to her directly, has guided her thoughts or lifted her spirits? The social scientist cannot claim that these responses are incompatible with her research, since she does not possess the sort of exceptionless laws that could rule out direct inspiration by God. She might well feel personally that such claims are highly unlikely, but this is a far cry from the principle of exclusion that rests on the success of lawlike explanation in natural science.

A common argument against miracles in the human realm is based on the claim that all mental phenomena will at some time in the future be reduced to physical phenomena, say to neurophysiological states of the body. According to the 'reductionist' programme in cognitive science (e.g. Churchland, Armstrong, Stitch), there is a mind–brain identity: every mental state corresponds to a brain state, and every mental operation shows a one-to-one correspondence with a set of neurological occurrences (see Chapter 8 below). Should this reductionist programme turn out to be successful at some point in the future, should such a one-to-one correspondence be established – but not before – the fate of miracles in the human realm will be the same as it is in the natural realm. In that case, the presumption of naturalism would be just as strong in both fields, and we should cease to make knowledge claims about psychological miracles. At present, however, it is hotly disputed whether the realm of the mental can ever be reduced to the physical; certainly such a reduction has not yet been accomplished. The only way to predict that someday it will be is to make a *metaphysical* assumption, the assumption of metaphysical naturalism or materialism. Since the reductionist programme is at present no more than a promissory note, it is a mistake to appeal to the ideal itself as evidence of its actual possibility. Identity theorists may hope one day to be successful, but hope is not empirical evidence. Given the present state of knowledge, given that we just don't understand the complexity of mental causation as well as we understand the laws of physical causation, the possibility of psychological miracles remains fully open. Consequently, there is less onus on the Christian when claiming that God has spoken directly to her in her mind than she faces when she claims that God has caused some specific physical event. At present, there may be knowledge of psychological miracles.

It could be objected, of course, that the social sciences consist entirely of 'judgements' about the actions and motivations of persons and groups. When I decide that Janice is angry with me, that Bill is prejudiced or that Frank's wife has ceased to love him, I do not base my judgement on general laws that predict human behaviours without exceptions. Psychology is, almost by definition, idiosyncratic; it describes the behaviour of individual agents without subsuming their behaviours to instances of universal laws. In part this is because the realm of the mental is one to which we have only indirect access (with the possible exception of the 'private knowledge' available to the individual). I can judge a particular action to be consistent with the way in which an individual normally behaves, or I can judge the action to be an anomaly in terms of what we would have expected, but I cannot say that a law has been broken when I see a strange behaviour. We may choose to use the word 'knowledge' when referring to our judgements in the human realm. But if we do so, we must at least be clear that it is knowledge in a very different sense from what we know and the way we know in the realm of the natural sciences. If, in speaking of our judgements of human behaviours and intentions, we use the word 'knowledge', then there *may* be knowledge of divine intentions and actions in the minds of individual persons. If, by contrast, we choose to speak only of *judgements* in the social sciences, then we should conclude that individuals may sometimes be able to make justified judgements that God has acted in this manner.

This analysis of judgements about intentions underscores, once again, why it is incorrect to speak of knowing that a physical miracle has occurred. When I judge that God has been responsible, partly or in whole, for a particular event in the world, I am making a judgement about the actions – and therefore about the goals and intentions – of an *agent* who is understood to be, in some significant sense, analogous to a human agent. That is, I am making the sort of judgement that one normally makes in the social sciences. This is one reason why miracles have traditionally been held to represent incursions into the natural order: they involve employing a *different kind* of explanation – an explanation in terms of purposes and goals – in a realm where purposes and goals are not (or have not been, at least since the advent of modern science) employed in explanations. Put differently: when someone claims that God accomplished a particular action, she is saying that the event cannot be explained in the way in which we normally have knowledge of natural events. Instead, she judges that this particular event is much more like an action accomplished by a human actor, as when I say that the glass of water was spilled because I knocked it over in order to express my anger. It is because miracle claims are of this 'mixed type' that we must give them a different epistemic status than we give to those natural events which we ascribe to the blind workings of natural law.

CONCLUSION

We have found that the presumption in favour of scientific explanations in the natural realm and an irreducibly subjective component in the case for any specific miracle keep us from using 'knowledge' of traditional miracle claims. Against the scientific naturalist or the Humean sceptic, we may conclude that there may sometimes be reasons to claim that God has acted; but against advocates of 'religious perception' such as William Alston I have argued that these reasons do not justify full claims to knowledge.

The interaction between more subjective or judgement-based reasons and knowledge-based reasons is a complex one and may move in both directions. A religious judgement might justifiably lead me to reject certain historical arguments – say, the argument of Dominic Crossan that Jesus' body was probably eaten by dogs.[16] But it might also be affected by developments in historiography, for example by the discovery that none of the gospels was written before the year 100 CE. This interplay should not be surprising: academic discussions are not 'pure'; they rely on judgements (e.g. the judgement about a historical reconstruction, 'But people just don't act like that!'). There is an intricate interplay between propositional and judgement-type claims, and different cultures may form different judgements about what humans would or would not do. And yet reason is not otiose: beliefs do become more justified when they are based on careful reasoning that has been subjected to criticism and revision at the hands of other experts in the field in question.

The role of judgement and interpretation becomes obvious when there is an actor, especially one outside the natural order. I judge that you intended that last comment as a joke, based on a certain facial expression you made and my knowledge of how you tend to react in situations like this one; but I also acknowledge that my judgement represents an interpretation of your probable intention and that further evidence might cause me to revise it. Similarly, the believer might judge that the behaviour of the disciples soon after Jesus' death is best explained by God's having intervened miraculously in history, raising Jesus from the dead and causing him to be seen by various witnesses. She might make this judgement *despite* her presumption that events in the world are best explained by natural regularities, judging that in this particular case factors inaccessible to the scientist *qua* scientist overcome the presumption of naturalism. Because these other factors are more subjective and not accessible to criticism in the same way that scientific explanations are, she does not claim for her judgement the status of knowledge. It may be a paradoxical occurrence that she has come to believe in, in so far as it involves the claim that a (divine) agent has acted intentionally in a manner in tension with the natural order for which science is normally responsible. Yet this paradox – appealing to the action of an agent where we would normally appeal only to the regularities of the natural order – does not represent a metaphysical impossibility, and a person might sometimes have good reason to believe that a supernatural event has occurred.

THEOLOGICAL IMPLICATIONS

The results of our study here have been neither as helpful nor as destructive to theology as is often alleged. On the one hand, most theologians who will be reading this book already find themselves in a context in which the sciences have the first (and, unfortunately, often the last!) word regarding knowledge of natural phenomena. I have argued that this fact makes it problematic for one to claim knowledge of miracles in the traditional sense. On the other hand, the standard ('Humean') rejection of divine interventions as impossible turns out to be unjustified. Just as we found in Chapter 5 that scientific cosmology does not invalidate talk of a divine cause, and may actually suggest it, we now find that careful conceptual analysis does not, as is often claimed, 'prove the impossibility of divine interventions'. Analysis does show, however, that talk of miracles involves interpreting the natural world not in terms of objects and forces alone, but ultimately in terms of the actions of a divine agent.

We have reached a major parting of the ways between theology and its naturalist discussion partners, the importance of which cannot be overemphasised. It *is* possible to give the last word to the sort of regular, lawlike knowledge that we have of the physical world. The proclamation of Christian theology, however, is that the world which evidences these lawlike regularities is *also* to be interpreted in terms of underlying intentions and purposes. We have found that this *personalist hypothesis* about the universe is a judgement similar to the judgements that we make about the behaviour of other persons. As such it cannot be assessed in terms of scientific laws alone; it demands to be treated in terms appropriate to the actions and intentions of agents. (Of course, it is also more controversial than the beliefs we hold about other human agents, since in this case *the existence of the agent himself* is a matter of deep dispute among the discussion partners in the debate.)

This conclusion raises two further questions of its own. First, what does it mean to ascribe *any* actions in the world to a divine agent? The theology of panentheism that we have developed suggests that God is related to the world in a sense analogous to the relation of our minds to our bodies. But how, in light of what science now tells us about the world, are we to understand the claim that God brings about changes in the world? The contribution of quantum and chaos theory to a theological theory of divine agency will occupy us in the next chapter. Later we will step back a bit from the specifics and attempt, in light of them, a broader theological account of God's relation to the world that does justice both to these scientific parameters and to the theological parameters established in earlier chapters.

Second and finally, *who* is it who acts in this way? As we have seen, the Christian tradition teaches some very specific things about the God who acts in the world, and particularly about God's salvific actions in Jesus Christ. We will find that the scientific and philosophical parameters on divine action that we have explored will cause us to understand that tradition differently in some respects. Conversely, however, the content of the Christian tradition also fills and transforms those parameters, allowing them to appear in a uniquely Christian theological context. It is appropriate that we complete our study in the final chapter by tracing this transformation.

NOTES

1. Owen C. Thomas (ed.), *God's Activity in the World: The Contemporary Problem* (Chico, CA: Scholars Press, 1983).
2. In Thomas (ed.), *God's Activity in the World*, p. 78.
3. There are several recent collections of articles on naturalism that the interested reader might consult. These include Peter A. French, T. E. Uehling, Jr and H. K. Wettstein (ed.), *Philosophical Naturalism*, Midwest Studies in Philosophy, Vol. 19 (Notre Dame, IN: University of Notre Dame Press, 1994), and Steven J. Wagner and Richard Warner (eds), *Naturalism: A Critical Appraisal* (Notre Dame, IN: University of Notre Dame Press, 1993).
4. See Thomas F. Tracy (ed.), *The God Who Acts: Philosophical and Theological Explorations* (University Park: Pennsylvania State University Press, 1994). See also Diogenes Allen, *Christian Belief in a Postmodern World* (Louisville, KY: Westminster/John Knox, 1989), especially Chapter 9, 'Divine agency in a scientific world'.
5. See Thomas V. Morris (ed.), *Divine and Human Action: Essays in the Metaphysics of Theism* (Ithaca, NY: Cornell University Press, 1988).
6. Wolterstorff, *Divine Discourse* (Cambridge: Cambridge University Press, 1994), pp. 273ff.
7. E. E. Evans-Pritchard, *Nuer Religion* (Oxford: Clarendon Press, 1962).
8. See William P. Alston, *Perceiving God: The Epistemology of Religious Experience* (Ithaca, NY: Cornell University Press, 1991). Alvin Plantinga and Nicholas Wolterstorff have both espoused the Parity Thesis to one degree or another; see Plantinga's ground-breaking article in Plantinga and Wolterstorff (eds), *Faith and Rationality: Reason and Belief in God* (Notre Dame, IN: University of Notre Dame Press, 1983), and Wolterstorff, *Divine Discourse*. The Parity Thesis has been criticised, *inter alia*, by Mark McLeod, *Rationality and Theistic Belief: An Essay on Reformed Epistemology* (Ithaca, NY: Cornell University Press, 1993).

9. See David Hume, *Dialogues Concerning Natural Religion*, ed. Richard Popkin (Indianapolis: Hackett, 1980), Part 2.

10. See Wolfhart Pannenberg, *Revelation as History*, trans. David Granskou (New York: Macmillan, 1968).

11. See Austin Farrer, *Faith and Speculation* (New York: New York University Press, 1967), and Edward Henderson and Brian Hebblethwaite (eds), *Divine Action: Studies Inspired by the Philosophical Theology of Austin Farrer* (Edinburgh: T. & T. Clark, 1990).

12. Thomas Aquinas seems to have had something similar in mind when he spoke of the efficient physical causes of every event, and yet simultaneously maintained God's role as a final or formal cause of every event; see *Summa Theologiae*, I/1, q. 44, art. 4.

13. Tracy (ed.), *The God Who Acts*.

14. See Alston, *Perceiving God*; Wolterstorff, *Divine Discourse*.

15. See the work of Steven T. Davis, e.g. 'Is it possible to know that Jesus was raised from the dead', *Faith and Philosophy* 1 (1984): 147–59; the sceptical critique by James A. Keller, 'Contemporary Christian doubts about the resurrection', *Faith and Philosophy* 5 (1988): 40–60; and the rejoinder by Davis, 'Doubting the resurrection: A reply to James A. Keller', *Faith and Philosophy* 7 (1990): 112–16. More recently see Davis, *Risen Indeed: Making Sense of the Resurrection* (Grand Rapids, MI: Eerdmans, 1993).

16. See John Dominic Crossan, *The Historical Jesus: The Life of a Mediterranean Jewish Peasant* (San Francisco: Harper, 1991); Crossan, *Who Killed Jesus? Exposing the Roots of Anti-Semitism in the Gospel Story of the Death of Jesus* (San Francisco: Harper, 1995).

7

SCIENTIFIC CAUSALITY, DIVINE CAUSALITY

—⟶∿∿∿ᎡᎡᎡᎡ⦿ᎡᎡᎡᎡᎡ∿∿⟵—

DIVINE ACTION IN AN AGE OF SCIENCE

We have seen that the history of theism has developed in two major stages. First came the development of what we have called radical monotheism out of polytheism. Later the question of the relationship of this God to the world had to be thought through again in light of the unmistakable successes of science. Unfortunately, in the process the equation of pantheism with atheism and materialism led to a suspicion of, even hostility toward, all views that drew a close link between God and the world. This suspicion delayed the development of panentheism for several centuries, as evidenced in the overreaction of Western thinkers against the pantheism of Spinoza's metaphysics. In recent years, as we saw, the strengths of panentheism – its ability to satisfy both the biblical revelation and the demands of philosophy – have come to the fore.

Along with the question of God's ontological relation to the world comes the question of God's activity in the world. The latter topic, like the debate between classical theism and panentheism, boils down to a fundamental distinction within the theistic camp: that between theism and deism. All agree that it will not be enough for traditional Christian theology if God's only role is to initiate the creation of the world in the beginning but to have no influence on it after that point. Yet many are drawn to (forced to?) deism, the view that God has no further causal influence on the world beyond creation, because of the difficulty of conceiving actual causal incursions ('interventions') of God into the world. As we have seen, not only science but also some theological arguments tell against such a model. Clearly, it is an urgent task for the theologian to provide a clear account of what she means by asserting that God continues to be active in the world. As we have seen, the urgency does not stem from a desire to do natural theology in opposition to theologians such as Barth, who wish to focus on the logic of God's self-revelation. Instead, it is the necessity of *understanding* what it is that God has done in revealing himself that leads theologians to address this question in dialogue with philosophy and the sciences. The integration of theology and science is as much a task of theology proper as it is an apologetic function.

The Changed Context for Divine Action

The present-day crisis in the notion of divine action has resulted as much as anything from a shift in the notion of causation. In pre-mechanistic science, science dominated by the influence of Aristotle, a component of divine causal action or teleology was included in every action. Thus Thomas Aquinas insisted that every event involved not only the efficient cause (what we would now speak of as *the* cause of an occurrence), but also the formal and material causes, or the influence of the matter and the form on the outcome. As a fourth type of causality, Thomas stressed the role of 'final causes', that is, the overall purposes of God, which act as one of the causal forces in every event.

It is true that some contemporary theologians have attempted to preserve something like this 'final' type of causality. One of the most sophisticated representatives is Wolfhart Pannenberg. In chapter 4 of *Theology and the Kingdom of God*, he adopts something like Aristotelian final causality, speaking of the power of the future as a causal constituent in every event.[1] A similar adaptation or version of final causality is visible in Lewis Ford's 'lure of the future', a notion that he adapts from Whitehead. Thomistic overtones can also be heard in theories of divine action that distinguish between primary and secondary causality – indirectly in the work of Austin Farrer and more directly in the writings of David Burrell.[2]

Such defences of 'future causality' in one guise or another cannot be quickly dismissed as metaphysical non-starters.[3] Nonetheless, one must still ask whether a *primary focus* on God's causal activity as emerging from the future can do justice to the changed way of thinking about causality that has dominated scientific theories of the world for the last three hundred years. When a full explanation of a physical event is given in terms of the causal forces that brought it about (efficient causality), this is generally taken as an adequate explanation of the event. We do speak differently of actions carried out by intentional agents, in so far as the purposes that the agent has in mind belong among the causal determinants of her actions over an extended period of time (say, as she works to complete her degree). But the 'intentionalist' model is not generally applied in explaining events in the physical world. That is, we do not take it that, in addition to the explanation of a physical event in terms of physical causes, it is also necessary to explain it in terms of cosmic purposes. One may wish eventually to challenge this widespread tendency, but one can only do so once one has understood how strong its hold is on modern thought.

The shift in thinking just described creates a completely different context for theological reflection on divine action. More precisely, it forces on the theologian a new set of alternatives. It would now appear, at least at first blush, that *either* God acts as the Divine Architect, who created a finely tuned machine and left it to function in a perfect manner expressive of its Designer, *or* God becomes the Divine Repairman, whose imperfect building of the machine in the first place requires him, like an inept refrigerator repairman, to return from time to time to fix up errors he made the first time around. Though perhaps not impossible, it is certainly difficult to recreate the pre-scientific framework – or to develop an alternative one – that allows one, alongside the network of scientific explanations, to speak of another causal system which is equally constitutive of the events in question.

Easy Interventions, Difficult Interventions

It is not difficult to construct a continuum of types of intervention.[4] For Christian theism, the easiest type of divine activity to introduce is creation. Contemporary cosmology allows us to think of some source for the Big Bang. Since that source exists at a moment prior to space and time as we know them, by definition no physical constraints can be placed upon it. (This assumes that the Big Bang is a 'singularity', an event that is not an instantiation of any more all-encompassing physical laws.) Thus it is fully open to theology to speak of a divine cause of all that is. Although it is clear that theologians will want to preserve this role for God within the doctrine of creation, most maintain that it alone does not suffice for a distinctively Christian theism. Deists, for example, are perfectly happy to accept this type of divine action.

The next easiest type of divine involvement to maintain is 'conservation', or God's role in sustaining the universe. Since the notion of sustaining is a metaphysical notion, it does not interfere with any particular natural scientific account. No natural law or natural explanation is changed or challenged when we add, 'and making all of these interactions possible is the continual will of God that the universe continue in existence.' An atheist might object that things in the universe can continue perfectly well on their own without the need for outside assistance, but this too is a metaphysical statement and would have to be defended as such. Divine conservation is not in itself a problematic contention for the scientist. If, indeed, as Jesus says, God 'causes his sun to rise on the evil and the good, and sends rain on the righteous and the unrighteous' (Matt. 5: 45), no conflict with science need be envisioned.

The next easiest type of intervention to talk about is a psychological intervention. It is easier to maintain that God 'brought' the needy individual to the church service than it is to argue that God fixed the furnace apart from any human agency. The reason for the difference is that we do not possess (at present) laws of human behaviour; social scientists can ascertain at most broad patterns of human response, and even these admit an amazing variety of (personal and cultural) exceptions. In the human realm, uniqueness and idiosyncracy are still the norm. Thus no laws are broken when we speak of an individual action in a non-standard way – indeed, this is almost what we *mean* by an individual action! 'Psychological miracles', as we saw in the previous chapter, could be frequent occurrences; God could often guide and direct the thoughts and feelings of individuals without causing an affront to natural scientific knowledge of the world. This is why one finds so little resistance to a 'psychological' account of Jesus' resurrection. I recall, for example, listening to various conversations within the Jesus Seminar – a group famous for its resistance to supernatural miracles in the New Testament documents – in which the reawakened faith and hope on the part of the disciples was gladly designated as 'the resurrection of the Christ'. Christian theologians across the liberal/conservative spectrum share an emphasis on the presence of God in individuals' personal experience and on God's providential role in guiding them 'toward all truth' (John 16: 13; cf. 14: 17, 26; 15: 26).

Matters are very different in the natural realm, however. Here we possess an amazing variety of laws and powerful explanations of individual occurrences. We have already seen that it would be metaphysical prejudice to rule out any chance of direct divine action in the natural world; still, the evidence is on the side of regularity. Theological claims for divine action in the natural world are much more difficult to maintain than those that talk about God's influencing the thoughts, wishes or decisions of an individual person.

The New Theological Task

To answer this crucial challenge, then, we must enter into the debate that has taken place over the last few years between theologians and scientists. If one is to offer a full theory of divine agency, one must include some account of where the 'causal joint' is at which God's action directly impacts on the world. To do this requires one in turn to get one's hands dirty with the actual scientific data and theories, including the basic features of relativity theory, quantum mechanics and (more recently) chaos theory. Fortunately, we do not need to begin at ground zero, for a number of theologians have been addressing these very questions in a high-profile debate over the last few years. Prominent among the forums has been the ongoing project of the Vatican Observatory and the Berkeley Center for Theology and the Natural Sciences, reflected in a five-volume series of books on the central questions of theology and science. We thus join the debate midstream, bringing it to bear directly on the theological issues that we have been examining in the preceding pages.

For the scientist-theologians writing in this area, the most pressing issue is how to avoid de facto deism – not merely by calling it unorthodox and expressing their dislike of it, but by actually showing why it is an unnecessary conclusion to draw from contemporary science. Most believe that Christian theologians must be in the position to say what they *mean* by God's activity in the world and *how* God's activity can be consistent with the belief that God has created a finite order with a goodness and a perfection of its own. If theologians fail at this task, they have failed not only in apologetics; they have also failed to make sense of their own views. The inability to produce a theory of divine action would represent an *internal* (and thus a theological) difficulty for Christianity, since the task has to do with meeting one's own standards. (Perhaps this is why this topic has driven more theologians toward deism than have any 'external' criticisms of Christianity.) Let us see, then, what kind of progress can be made toward giving a positive account of God's agency in the world – one that preserves the divinely established order in creation and works with, rather than against, the science that reflects that order.

FINDING GENUINE OPENNESS IN THE PHYSICAL WORLD

Some of the most important recent approaches to thinking of God's causal activity in the physical world make use of the physics of the smallest particles, quantum mechanics. This is certainly a natural area for scientifically minded theologians to look since, as the physicist Karl Young notes, 'All current bets in physics are on a fundamental theory of the natural world being based essentially on relativity and quantum mechanics.'[5] Remember that the question is not how to *prove* that God is active in the world at particular moments, but rather how to think this possibility in a manner that does not conflict with what we now know of the world. If theologians were to argue that some area of the physical world can never be understood by science because it involves supernatural causes inaccessible even in principle to scientists, then they would be setting a priori limits on what physics can ever establish – limits motivated not by scientific developments but by theological arguments. History has shown such arguments to be extremely vulnerable to falsification. Equally dangerous is to situate God's causal activity in the 'cracks' between existing scientific theories. Such approaches are correctly accused of appealing to a 'God of the gaps', since they make theology dependent on what science does not yet know. Time and time again in the history of science new theories have closed these gaps, leaving Christians high and dry in their attempt to speak of the action of God. Such failures, repeated over and over again, have understandably led to increasing scepticism about whether a place can ever be found for divine causal activity.

The Answer of Quantum Mechanics

This is the reason why quantum physics seems to offer such a hopeful arena. For the most widely accepted interpretation of the equations of quantum probabilities holds that the limitations on our knowledge of quantum states are intrinsic to the world itself. It is not that better theories or measuring apparatus will someday allow us to specify the location and momentum of a subatomic particle with exact precision. Rather, there are good physical reasons to think that precise knowledge of this sort will never be possible (the Heisenberg uncertainty principle), in part because the amount of energy necessary for producing this measurement would change or eliminate the very state we were seeking to measure. More controversial but still widely shared is the so-called

Copenhagen interpretation of quantum mechanics, which argues for an actual *ontological indeterminacy*: nature itself is indeterminate at the quantum level. Imagine a vacuum container containing nothing except a single electron, and imagine asking whether, at a given moment, that electron is in the right or the left half of the container. It is not just that *we* could never know which answer is correct; rather, nature itself is indeterminate on this question (until we measure it, at which time we will 'locate' it on one side or the other).[6] Similarly, we can specify the average half-life of uranium-235 atoms, but we believe it to be impossible, even in principle, to predict when one particular radioactive atom will decay.

There is no other area within physics that reveals this sort of ontological limitation built right into the structure of the existing universe. (As we discovered above, moments before the Big Bang are physically unknowable. But such moments are *outside* the physical universe itself and thus not surprisingly unknowable.) It is for this reason that quantum mechanics suggests itself so strongly as an area in which sense could be made of divine intervention *without* doing damage to the integrity of science *or* making oneself vulnerable to the embarrassment of further scientific progress. Thus the physicist-theologian Robert J. Russell, who has had a crucial influence on a number of the authors cited in this chapter, has been a leading advocate of the view that God could intervene supernaturally within the scope of quantum indeterminacy. Given billions and billions of such minute interventions – the potential number would be limited neither by science nor by inability on the part of God – God *might* be able to effect significant changes on the macroscopic level.

There is another feature of the quantum approach that is attractive to the theologian: scientists would not necessarily be able to detect any particular pattern at the microscopic level, even if God were continually guiding the universe by this means. This fact is important because thinkers have worried in the past that, if a particular locus for God's action were specified, humanity would then possess a proof of God's existence and action. It would follow, they worried, from a pattern of divine guidance emerging unambiguously *within the discipline of physics itself*, that the role of religious faith would disappear and 'God' would become merely a component of natural science itself – certainly a *reductio ad absurdum* of natural theology! Yet this theologically depressing result could never occur in quantum physics, since we now believe that there *can be* no laws that predict quantum outcomes beyond the probability functions. Interestingly, however, the pattern of divine guidance might well become visible at a 'higher' level, for instance when one steps back from quantum physics and begins to look at, say, the pattern of history as a whole. But that is how it ought to be, the theologian will respond: patterns of divine intervention *should* be fully recognisable only in theological terms, or at least on a level of activity that is appropriate to God as actor.

Other Proposals for Locating Divine Intervention within Physics

Several other proposals have been made for showing the consonance of divine action and physical theory. Most widely discussed of late is the impact of recent work on 'chaos theory', or the physics of dynamic systems far from thermodynamic equilibrium. The physicist John Polkinghorne, for example, has suggested that the phenomenon of chaos, whereby tiny changes in initial conditions have future causal effects that are essentially unpredictable, might represent an 'intrinsic indefiniteness' or indeterminism within (at least parts of) the *macrophysical* world.[7] Clearly, in chaotic systems future states of the universe cannot be predicted from present states, since differences in the initial conditions that are too small for us to detect will lead to vastly different future states of the system. (We see this all the time in weather forecasting, where small differences in the cloud and air masses as they proceed toward a given location lead to vast differences in the actual weather conditions a few days later – differences between, for example, a rainy day or a warm sunshiny one.) If Polkinghorne is right, there might be genuine openness (indeterminacy) not only at the quantum level but also in the 'chaotic' parts of the macrophysical world. If so, God *might* act in these systems to bring about divine intentions without breaking any physical laws ('interventionism').

There seem to be some difficulties with this suggestion, however. For example, are chaotic systems really 'indeterminate' or causally open? Most physicists treat them as causally determinate systems. Quantum physics may accept actual or ontological indeterminacy at the subatomic level if (as many believe) there is no single state of affairs or hidden variable 'underneath' the quantum probability functions. By contrast, chaos theory seems to presuppose that the initial conditions are actually one way rather than another. If this is true, the indeterminacy will lie only in the degree of *our* knowledge of these conditions; the inference from epistemic limit to ontological limit will be blocked. Once the initial conditions are determinately given (and known exhaustively by an omniscient being), then the amplification effects studied by chaos theorists function in a purely deterministic manner. Perhaps God, who could presumably know the initial states of some system with infinite accuracy, could cause changes at the quantum level and then use the famous 'amplification effects' of chaotic systems to amplify these changes into macrophysical outcomes – say, to make it rain on the just and unjust alike. But if chaos theory does turn out to be a subset of deterministic physics, then the attempt to use it not just as an amplification device but also as an actual opening for divine action would turn out to be another 'God of the gaps' strategy. Such are the potential gains, but also the risks, of cutting-edge work in theology and science.

The Default Position

Should none of the proposals we are about to consider pan out, the theologian will be left with two major options. On the one hand, she can give up her quest to specify the 'causal joint' where divine action occurs and yet still maintain that God is active in the world in *some* way. Perhaps she would wish to assert that, although God does not carry out any specific actions in the world subsequent to creation, the continued existence of the universe should be understood as a single composite act of God, as Maurice Wiles does in *God's Action in the World*.[8] Or she might speak, following Austin Farrer, of a 'primary causality' of God, in and behind every action carried out by every finite agent. This route, similar to the view which Nancy Murphy calls 'immanentism' and associates with modern liberal theology,[9] has the advantage of making God causally ubiquitous in the world – though it might also seem that the God who does everything is a God who does nothing. Or she might designate it a matter of pure faith how God brings about his purposes in the world, beyond all grasp of human reason. On the other hand, the theologian might give up altogether on the claim that God acts in the world. Such a move appears also to move outside of Christian theism, at least in the sense in which the tradition has understood it, siding instead with deism, pantheism or atheism.

Clearly, then, resolving the 'causal joint' debate is not the *sine qua non* for the survival of Christian theism, although asserting divine action in *some* sense may be. Nonetheless, it should be clear that it is a high-stakes debate for those who wish to pursue Christian theology in dialogue with the sciences and our general knowledge of the world.

POSSIBLE INTERVENTIONS AND ACTUAL INTERVENTIONS

As an opening example of what is involved in this debate, we might consider a recent collection of articles entitled *The God Who Acts*, many of which argue for the possibility of divine actions in the world.[10] Again, it is important to note that a positive answer to this question does not yet address the *epistemological* question of whether one can know that God acts supernaturally within the world. For one can imagine a scientifically-minded critic responding, thus:

> Sure, I admit the *theoretical* possibility of a 'force' affecting the world whose action is not empirically measurable in any way. Perhaps this force resolves quantum indeterminacies in such a way that no broader patterns are detectable to observers of quantum events. The force might even use the magnifying effects of chaos theory (which are also undetectable and unpredictable) to bring about certain macrophysical outcomes. If the force is conscious and has intentions, then it might even actualise its intentions by these means – say, preserving the life of one individual or bringing about the death of another. You theologians have made the case that such interventions *could* take place without breaking any physical laws, hence I grant that there *could be* divine action in the world. However, what actual reasons do you have for thinking that such events actually occur? You may speak from the perspective of faith, if you wish, about what purposes God has in the world. But the purposes of such a being, not to mention its very existence, are by your own admission in principle inaccessible to empirical study. Thus I disbelieve in your hypothesis – not because it is impossible, but because there is no serious evidence for it at all.

In the collection of articles mentioned above, the opening piece by Maurice Wiles, 'Divine action: Some moral considerations', serves as a sort of whipping post for most of the subsequent contributions, most of which distance themselves from Wiles in some way. In light of the imagined criticism just formulated, one wonders whether they have really understood what motivates Wiles to his conclusions. Wiles does in fact bring a variety of *epistemic* arguments as part of his case against specific divine actions in the world. Thus, for example, he alludes to the problem of evil: the presumption that God must be responsible for the evil that occurs in the world, and for the suffering that goes unprevented and unrevenged, if he is said to be capable of preventing it and yet does not.

That God is capable would follow immediately from the fact that he does in fact sometimes intervene to prevent suffering (pp. 16f.). Wiles also employs arguments that tell against the possibility in principle of direct or special acts of God (pp. 22f.). But none of the other authors seem to recognise that one of the important, if not the most important, of Wiles' arguments against special divine interventions is the epistemic argument. Wiles writes:

> Rather, God is our symbol for that which underlies, gives existence to, and makes sense of our finite and temporal existence. His 'eternity' is only known to us as the ground of that which has a beginning. Our puzzlement about how an eternal God can create the beginning is not like our puzzlement about the magnet [and how it makes iron filings jump]. To treat it as if it were can generate a style of discussion of divine attributes (as truths about God known independently of any knowledge about the world) that is inconsistent with the way in which the concept of God is grounded. (pp. 14f.).

Wiles is concerned with the way in which the term God is introduced in the first place. He stresses that it is a metaphysical or theological term, one which refers to the ground of all finite things, the source of the finite world, the purpose that, believers say, exists in and behind it. How, he asks, could one begin to ground the claim that the source-of-all-things is the specific causal agent responsible for bringing about a particular state of affairs within the world? Many of Wiles' critics in the book work to show – and I think successfully – that such an identification is not the metaphysical impossibility that Wiles seems to suggest. But what about the *other* contention, that we would never have adequate reasons for identifying the God of all-that-is as the cause of some particular event? If one cannot even imagine how one would go about justifying such a claim – and even more if one possesses some evidence that God does not act when he might be expected to (as the problem of evil suggests) – then, Wiles might argue, one is better off not asserting that God does actually act in concrete ways in the world, lest this assertion become arbitrary. It is for this sort of reason, I suggest, that Wiles finally advocates what he calls 'a highly revisionist thesis', namely the proposal

> that our inherited Christian language about particular acts of God is best understood, not in relation to divine initiation or direction, but of actions whose results further the overall intention of God in the creation of the world or, to put in more temporal terms, his

will for the world at a particular moment in the light of the way in which we have developed its potential up to that point. (p. 25).

One way to address Wiles' objection is to provide a way of understanding direct divine interventions, if such do occur. The problem that one immediately runs into is the problem of human freedom: if God is responsible for *directly* bringing about certain states of affairs, then human actors cannot be responsible for the outcomes. When outcomes are determined, human freedom disappears; we seem to become puppets on a stage whose every action has God as its ultimate cause. Such a picture conflicts with the fundamental theological notion of human responsibility before God, which surely presupposes some sort of human freedom. Yet if humans are indeed the cause of their actions, how can one attribute those actions to God as well?

A number of solutions to this problem have been attempted. Austin Farrer, followed by Henderson, Hebblethwaite and others, argued for a notion of 'double agency'. There are two types of causes for any event: a string of finite causes within the world brings it about, and at the same time God is active in bringing about the outcome through a separate type of causality. Kathryn Tanner has presented a sophisticated version of this view:

> The theologian talks of an ordered nexus of created causes and effects in a relation of total and immediate dependence upon divine agency. Two different orders of efficacy become evident: along a 'horizontal' plane, an order of created causes and effects; along a 'vertical' plane, the order whereby God founds the former. Predicates applied to created beings . . . can be understood to hold simply within the horizontal plane of relations among created beings.[11]

Thomas Tracy resists, I think rightly, this version of double agency. He pushes the argument that humans either are or are not the cause of their actions; an action of mine is either a basic action, for which I am responsible, or an action for which God is responsible. Under this view, only if I am the cause of my action is my responsibility preserved.

Such a view of action implies that God's action in the world should be understood as something more like divine persuasion. Tracy concludes:

> There are, therefore, important respects in which the free acts of creatures can be regarded as God's acts. If we deny that God is the sufficient cause of the creature's free acts, we can immediately go on to affirm that God acts with the infinite resources of omnipotence to guide those choices by shaping the orienting conditions under

which they are made. In untraceably many, varied, and subtle ways, God continuously brings to bear the pressure of the divine purpose for us without simply displacing our purposes for ourselves. God's action goes before our own, preparing us (in spite of ourselves) for the unsurpassably great good that God has promised us. (pp. 101f.)

This view does weaken the sense in which God's causal agency contributes to the actions that humans perform in the world, at least in comparison to many classical views. Nonetheless, it can still be theologically sufficient. Seen in this way, God's role becomes that of one who prepares and persuades, rather than the one who 'brings about' human actions. This view is also conceptually much neater than the alternatives, since it attributes basic actions to humans alone. At the same time, it does continue to ascribe a crucial role to God in 'luring' humanity and encouraging certain types of actions. Under this model there must be genuine openness in history. Pannenberg, among others, has made clear that the result is that one cannot *know* in advance that God will bring about the divinely desired ends.[12] Still, one can know that if God is God these ends will be achieved, and the final state of affairs will be consistent with God's nature.

So much for the basic framework within which a more concrete theory of divine action might be developed. Is it possible to go beyond this point and specify how God could be more concretely involved in the affairs of this world?

THE CONTEMPORARY DEBATE ABOUT DIVINE CAUSALITY IN SCIENCE

One finds in the recent literature a rich variety of positions on science and divine action. The goal of an introductory text must be restricted to examining some representative positions; I will thus limit the discussion in what follows to five of the major approaches that characterise the recent debate. Observing their agreements and disagreements, their strengths and weaknesses, goes a long way towards specifying the theory of divine action that we have been seeking. It will also provide the framework for expanding 'the panentheistic analogy' in the final chapter.

God's Action as 'Top-down' Causality[13]

John Polkinghorne holds the view (also defended in Chapter 5 above) that all knowledge claims about the world presuppose certain underlying metaphysical commitments. These commitments, which form the backbone of any view of the world, can in his view be judged only by the pragmatic criterion of their consistency with observed phenomena. When we consider the question of divine agency, we must use this pragmatic standard of evaluation as well. Historically, most thinkers have utilised some form of analogy to human agency in order to conceive divine agency. Polkinghorne notes that the majority of scientists are 'critical realists'. According to him, this means that, in the course of looking for the optimum correspondence between epistemology and ontology (between our theories and reality), scientists use epistemic or theoretical models to defend ontological premises. For example, in the case of Heiseinberg's uncertainty principle in quantum theory, the majority view takes the limitations on *our* knowledge to reflect ontological indeterminacy as well (assuming that the 'many worlds' interpretation and Bohm's 'hidden waves'[14] are both incorrect). Polkinghorne proclaims that scientists are indeed justified in treating epistemic limitations as evidence of a particular ontology; his goal is to use this procedure to find a place for divine agency.

The standard science-based response to theology is to dismiss talk of divine agency, since 'there is no place for it'. Such a response presupposes the metaphysical doctrine of *physicalism* (all objects, agents and causes are physical in nature). But, as Polkinghorne shows, the pursuit of science does not require one to hold the metaphysical doctrine of physicalism; it is a choice – or a prejudice. Nor has the development of quantum theory offered solace to those who would reduce all states of affairs to one or more fundamental particles (atomism).

By contrast, the strongly theological response is to speak of God as a contributing cause to all events. This is the strategy taken, for instance, in Austin Farrer's theory of 'double agency', in which God is present as 'primary cause' in, with and under all instances of creaturely or 'secondary' causality.[15] Such a view has recently won support and continues to represent a popular theological response to the question of divine agency.[16] Polkinghorne (following Peacocke) sees correctly that this view gives no help in thinking *how* God acts in the world. Of course, this criticism should not be presented as one that would have troubled Farrer himself, since he was quite clear as to what his view did and did not achieve:

> Such being the value of the analogy [of God as the mind of the world], it is clear that it casts no light whatever on the mysterious causal joint between prime agency (the Creator's) and second agency (the creature's), a relation which it simply presupposes.

God's being the mind of the world does nothing towards identifying his action with that of an organic whole, to which the actions
of the cellular constituents are geared.[17]

Can we say more about divine agency than Farrer did? The most ele-
gant alternative, according to Polkinghorne, involves the integration of
holistic ideals with the reductionist method. In this way we can enjoy
the successes that reductionist explanations have historically bestowed,
yet without the baggage of a narrow and devaluing outlook. Specifically,
science can employ the method of reductionism – explanation in terms
of fundamental forces and particles – without endorsing a metaphysics
of reductionism. (Think of the example of biological explanations, which
can admit the connections back to physics through biochemistry and
physical biology, but which can also employ the language of organismic
behaviour and ecosystem influence. We return to the theory of emergence
in the next chapter.)

Polkinghorne's alternative to reductionist theories of causation
involves a reversal of physics' customary 'bottom-up' approach. He suggests that, in addition to traditional reductionist explanations, we should
also consider the possibility that *a system as a whole* could motivate or
cause isolated changes in its component parts. Polkinghorne uses the
phrase 'top-down causality' to signify this type of change. One specific
area that Polkinghorne feels may be a likely candidate for 'top-down'
description is the relation between consciousness and the body. God's
action in the world, Polkinghorne speculates, may be similar to the
relationship of the soul to the body (though he rejects panentheism).

Polkinghorne urges two qualifications of the 'top-down' concept if it
is to be applied to divine causation.

1. Talk of God's agency, supported by appeals to the principles of
   self-organisation, must be based on phenomena not readily
   describable by the 'bottom-up' approach. True 'top-down' causality
   would have to have a more 'open' and non-local character.
2. If 'top-down' causality is a legitimate mode of explanation for a
   given set of circumstances, application of 'bottom-up' models should
   be insufficient to completely describe the observed phenomena in
   this context. Such failures of explanation-from-below would be
   the primary evidence available to us of the existence of 'top-down'
   causation.

Polkinghorne is of the opinion that our subjective awareness – the
irreducibility of consciousness – is adequate to justify rejecting any
metaphysical proposals that disallow 'top-down' causation.

I fear, however, that deciding *in practice* when 'bottom-up' failures are severe enough to warrant top-down explanations will prove a more difficult task than Polkinghorne imagines. For cannot the 'bottom-up' reductionist always appeal to errors in the data, or to missing or hidden variables, or to inaccuracies in our theoretical assumptions – or simply to the primitive state of science today? Thus the present debate in the philosophy of mind between classical dualists and identity theorists (reductionists) seems little influenced by the current failure of the reductionist approach; reductionists merely argue that their programme will be vindicated in the future. Nonetheless, Polkinghorne is right to stress that both sides must eventually make their case from the explanatory strength of their method, be it holist or reductionist. For theology, this means that we must show how assertions of divine activity in the world help to make sense of the human experience *of* that world. Accomplishing this task will involve also speaking of *who* God is as agent and *how* God is related to the world as a whole, precisely the two broader themes treated in this work.

Within current physics, there are two main domains that resist 'bottom-up' description; in both of these a 'top-down' approach may be more appropriate.

1. As we have noted, quantum theory seems to leave room for God to affect the outcomes of physical processes. If quantum events are inherently indeterminate (merely probabilistic rather than determined), then God could influence their outcomes without breaking physical law. As Polkinghorne writes elsewhere, 'Will not God's power to act as the cause of uncaused quantum events (always cleverly respecting the statistical regularities which are reflections of his faithfulness) give him the chance to play a manipulative role in a scientifically regular world?'[18] No scientific laws are broken, and a series of quantum-level occurrences can have major effects in the world, as in Schrödinger's famous thought experiment in which the decay of a single atom causes a poison to be released which kills a cat.[19]

   Quantum approaches to divine action, to which we will return in a moment, do, however, prompt Polkinghorne's scepticism: it is unclear how quantum interventions would be 'amplified' to lead to macrophysical outcomes (sparing the life of the child), and the opportunities for intervention would be too episodic to constitute a theologically adequate theory of divine action.[20] Polkinghorne points out elsewhere that we don't yet know how quantum events would cause changes in the brain that might affect thought. ('Are

synaptic firings capable of being effectively quantum phenomena?'
he asks.) He also doubts that this approach can provide a vibrant
enough account to make sense of divine action in the world: 'We
are back with a struggling ghost inside a now somewhat rickety
machine.'[21]

2. Chaos theory represents a more promising line to pursue. Chaotic
   systems, or systems far from thermodynamic equilibrium, show
   incredible sensitivity to initial conditions, such that predictions of
   outcomes quickly become impossible.[22] The inherently unpredictable nature of these systems, writes Polkinghorne, should lead
   us to entertain the possibility that they are 'open and integrated'
   (p. 153): 'open' in so far as reductionist elements do not absolutely
   determine future events, which leaves room for other causal principles; and 'integrated' because these other causal principles would
   have to be holistic or 'top-down' in nature. In short, there is the
   possibility that non-local, causal influences may be at work in such
   systems – including influences such as 'minds upon bodies' or 'God
   upon creation' (p. 154). He might have cited the similar conclusion
   about chaos theory by Crutchfield et al. elsewhere in the same
   volume: 'Through amplification of small fluctuations it [nature]
   can provide natural systems with access to novelty.'[23]

Such physical considerations lead Polkinghorne to a new theology of
divine action. Eschewing the three ruling frameworks for theories of
divine action in the past – materialism (which 'implausibly devalues the
mental'), idealism (which 'implausibly devalues the physical') and dualism
(which 'has never succeeded in satisfactorily integrating the disjoint
realms of matter and mind') – he suggests a *complementarity of influences
on the physical world*. At least in chaotic systems, the world may be influenced both by 'energetic transactions', that is, by physical or efficient
causal processes, *and* by 'active information'. God acts in the latter way,
influencing the formation of 'dynamic patterns' or overall contexts
rather than by means of 'transactions of energy'. Part of the reason that
Polkinghorne wishes to speak of divine influence as operating solely at
the level of information or context is that he wishes to preserve the
purely spiritual nature of God:

> God ... is not embodied in the universe and there does not seem
> to be any reason why God's interaction with creation should not
> be purely in the form of active information. This would corre-
> spond to the divine nature being pure spirit and it would give a

unique character to divine agency. . . (God is not just an invisible cause among other causes.) (pp. 155f.)

Polkinghorne's approach to divine causality, in other words, though concerned with the physical implications of this or that theory, is carefully sensitive to theological constraints. These include the conditions that God's action, first, must be continuous rather than fitful, capricious or occasional; second, it must involve interaction with the physical realm rather than 'interventions' that negate or disregard that realm; and third, it must work with and not against natural laws, which are 'expressions of the faithful will of the Creator' (pp. 244f.), for only in this way can one avoid the picture of a 'capricious, celestial conjurer'.

Elsewhere Polkinghorne considers two other ways that theologians have thought about how God might influence the world besides quantum indeterminacy and chaos. One way involves an appeal to *synchronicity*, the fact that certain significant events just 'seem to happen' at the right time. Some suggest that 'miracles' such as the parting of the Red Sea, the star over Jerusalem and the stilling of the wind as Jesus spoke were completely natural events, which God managed to set up in advance to happen in just this way 'by a combination of foresight and ingenious prior fixing'.[24] But this answer, Polkinghorne maintains, is inadequate as it stands. It doesn't tell us *how* God manages to set these things up in advance, and it pictures a universe so determined that one wonders how the significance and freedom of human actions can still be preserved. Also, religious believers usually mean something more robust by God's answer to prayer than that, countless millennia ago, God structured the physical world in such a way that, in our day, a certain action would happen just after I would 'happen to' pray for it.

Polkinghorne's other response is the appeal to direct miracles, which are 'signs, insights into a deeper rationality than that normally perceptible by us' (p. 76). He finds it credible that events like the resurrection would be part of a new 'regime' that is not normally accessible to us. Note that this view is similar to the one defended by Pannenberg: miracles are not defined as breaks in the natural order but rather as the first appearances of the new type of regularity that will characterise the world at future times.[25]

Polkinghorne's criticisms of other positions are insightful and his own hints at a theory of divine action are intriguing. Still, it could not be said that Polkinghorne develops a theory of divine action in detail. What is clear is that human action in the world serves as an important model for his reflections on divine action. He writes at one point: 'If the physical world can accommodate human action, it is not clear that it is not open to divine action also.'[26] It is this stress on the analogies with human action that will prove most fruitful as we turn to the so-called philosophy of mind debate in the next chapter.

Does Chaos Theory Provide an Opening for Divine Action?

As much as Polkinghorne's article and related publications stressed the more metaphysical side of the theology/science discussion, a major piece by Robert Russell (founder of the Berkeley Center for Theology and the Natural Sciences) and Wesley Wildman (Professor of Theology and Science at Boston University) reveals the scientific care and caution that must be brought to the subject.[27] Russell and Wildman include an introduction to the mathematics of chaos theory, and they are cautious about drawing robust conclusions from the physical evidence. Still, this caution allows the conclusions to emerge if anything more clearly.

The prior question that seems to motivate this inquiry (as it does other pieces by Russell) is whether science even leaves a place for divine action. If Laplace were right, for example, and a strict determinism would allow us to predict all future states and retrodict all past ones, then there would truly be no place for divine action. Conversely, if the physical world is truly indeterminate at some points – which is what many quantum physicists claim about the microphysical phenomena – then it would be metaphysically possible for God to act without setting aside natural law.

The problem is that indeterminacy is a metaphysical claim. Strictly speaking, all that the physicist can say is whether a physical system is predictable or not. If we wish to speak of randomness, of the sort that would allow God to act without breaking physical laws, we would be justified in positing it in those areas where scientific prediction fails. But, as Russell and Wildman show, finding evidence of genuine randomness is more complicated than one might think. 'Strict' randomness implies that the unpredictability of a system is absolute; 'eventual' randomness implies that that the system will progress from a state of predictability to complete unpredictability in a finite (and predictable) number of iterations; and the 'absence of' randomness occurs where our predictive power is (potentially) absolute. Only in the last case, strict determinism, is divine action impossible without breaking natural law. Since many physicists hold that there are cases of strict randomness in quantum physics, this remains an area in which theologians may speak of divine action without contradicting physical laws.

206

What about the second major area of the current discussion, chaos theory? Russell and Wildman show why randomness in dynamic or 'chaotic' systems does not easily confirm either determinism or indeterminism. A chaotic system is defined in part by its extreme sensitivity to initial boundary conditions. But this means that, by the nature of the theory itself, we can never find a direct experimental test to prove that a system is chaotic, since we can never define a physical system's initial state to a degree that would not lead to divergence from our theoretical model within a finite (and typically small) number of iterations.

As a result, Russell and Wildman's treatment shows convincingly that chaos theory is both bad news and good news *vis-à-vis* the quest for possible loci of divine action. It is bad news to the extent that it supports the determinist agenda; as they write, 'We can say without hesitation that chaos in nature gives no evidence of any metaphysical openness in nature' (p. 82). Neither the famous 'butterfly effect' – so named because the flap of a butterfly's wings in China could, in principle, affect weather systems over California – nor the fact that such systems are 'open to their environment' in a way that suggests a causal influence of the whole (of a system) on its parts proves indeterminism. The mathematics of chaos is completely deterministic, and its results mirror the actions of real dynamical systems, if only approximately. This is a victory for determinism: random-seeming systems in fact can be expressed mathematically, albeit in the mathematics of chaos; hence the sphere of genuine (or 'strict') randomness in nature is narrowed. In fact, the success of mathematical models in modelling dynamic systems suggests that *other* random-seeming physical systems might also someday be modelled in a similar fashion.

At the same time, chaos theory is also good news for theologians. The theory says that very minor changes in initial conditions – even ones so small that we cannot detect them – can have major impact on future states of the system. This means that there are limits on how far the determinist agenda can be confirmed. Humans simply could not know if God were affecting the future by intervening to make extremely small changes in initial conditions (say, on the scale of the flap of a butterfly's wing thousands of miles away) and then allowing the laws of chaotic steps to amplify them into major results (rain in a drought-ridden sub-Saharan country). Our century has been packed with apparently inherent limits on human knowledge – at the large scale (the so-called moment of creation or '$t = 0$' problem), at the small scale (Heisenberg's uncertainty principle), in formal systems (Gödel's incompleteness theorem), and even in our knowledge of ourselves (the self-reference problem). As Russell and Wildman write, 'Chaos theory highlights an epistemic limit in the macro-world of dynamical systems, tethering the deterministic hypothesis even as it advances it' (p. 83).

There is also a more speculative argument on chaos and divine action, hinted at by Russell and Wildman but perhaps worthy of more explicit attention. The physicist Charles Misner argues somewhere that the sensitivity to initial conditions in chaotic systems is so great that, in order to make long-term predictions, one would have to specify the location and momentum of the particles to a number of decimal places that exceeds the number of atoms in the universe. This constraint would also apply to God if he is attempting to determine future states. But (virtually) infinite accuracy could not be obtained without reference to quantum-level effects. Now, if current physics is right, quantum outcomes *cannot* be determined in advance; they are essentially indeterminate in nature. Therefore, God *could not* act only once, at the creation of the universe, and thereby determine the whole history of the universe and the particular outcomes he desires. If this argument is right, theological reflection faces some hard choices. One can espouse deism – God acted at the beginning but never again within the universe – or some form of process thought – God set the whole thing in motion but does not really control the outcome of the whole process. But if one wants to maintain something like the classical notion of providence – God guides the process of history according to divine goals toward an out-come God wishes to bring about – *then one cannot avoid some doctrine of divine action in the world.* Physically speaking, God can bring about future states of affairs without breaking natural law (say, through quantum-level causal influences), *but God cannot do it without any post-creation causality at all.*

Russell and Wildman provide an excellent model of the care needed today when speaking of God's bringing about changes in the natural world. Developments in science offer important theological resources, but they also impose subtle and sophisticated constraints on theological formulations. Nonetheless, there are grounds for optimism: chaos theory 'does open a window of hope for speaking intelligibly about special, natural-law-conforming divine acts, and it is a window that seems to be impossible in principle to close' (p. 86).

A Theory of Divine Action based on Quantum Indeterminacy

We have already encountered Thomas Tracy's work in the collection he edited, *The God Who Acts.*[28] His extended essay on divine action and the sciences[29] has the advantage of helping theologians step back from and 'place' the entire discussion within the theological tradition.

As we have seen, science has created a challenge to theology by its remarkable ability to explain and predict natural phenomena. Any theological system that ignores the picture of the world painted by scientific results is certain to be regarded with suspicion. But science is often identified with determinism. In a purely deterministic universe there would be no room for God to work in the world except through the sort of 'miraculous intervention' that Hume – and many of his readers – found to be so unsupportable. Thus many, both inside and outside of theology, have abandoned any doctrine of divine action as inherently incompatible with the natural sciences. To overcome the impasse it is necessary to raise the broader theological issue: what *concepts of divine agency* would be consistent or inconsistent with the tenets of modern science? Tracy specifies three significant models of divine action that attempt to address in one way or another the problems raised by modern science: the deterministic model of Friedrich Schleiermacher and Gordon Kaufman, the 'gap'-free model of Brian Hebblethwaite and John Compton, and his own response, a non-determinist, gap-dependent model.

Schleiermacher's model is a form of *dual-action theory* in which observed physical causation is understood as secondary to God's omnipresent and primary support of the world.[30] In this model, secondary deterministic causes are observed to be totally consistent and complete; there are no gaps. Schleiermacher holds these two different modes of causation completely distinct: one may speak *either* in terms of purely determinist physical causation *or* in terms of God's ultimate support and plan. The theological advantage of this model lies in the fact that the world may be viewed as a 'nature system' (p. 296); there need be no discord between science, history and theology. Unfortunately, there can be no direct interaction between God and humanity, no interventions of God within history once it is underway. Otherwise, the secondary causal structure would be broken and human knowers would be confronted with an 'uncaused cause' that would disrupt the perfect unfolding of history. The Christian must therefore grant that Jesus' role in history is determined: he can be at most a pre-ordained (or a self-motivated) realisation of the inherent human possibility of an 'unimpeded god-consciousness'(p. 298). It follows that God is only *indirectly* accessible through Jesus.

Tracy urges that this framework is not finally enough for Christian theology. For example, 'There can be no opposition between the divine and human wills' (p. 299): God would not *allow* pain and suffering but would have *caused* it through the creation of a predetermined history. In a criticism aimed more at Kant's attempt to preserve both pure reason, characterised by determinism, and practical reason, characterised by the assumption of human free will, Tracy concludes that Schleiermacher's acceptance of physical determinism must imply the non-existence of free will. Either our belief in free will is a fiction, or God must have created the world in anticipation of each of our free acts. Only a genuine openness in the physical world (physical indeterminism) would leave a place for genuine free will; but then it would, in principle, also leave a place for divine actions as well.

Kaufman's model retains the essential character of Schleiermacher's solution, though Tracy reads it as an unsuccessful attempt at refinement. (Similar efforts are found in the work of Maurice Wiles and Schubert Ogden.[31]) Kaufman agrees with Schleiermacher that God does not act directly in history after the moment of creation.[32] He wishes to speak instead of God's *master act*, 'the whole course of history'. Is *every* event an indirect expression of that master act, or are there divine 'sub-acts' that God uses to bring about the one master act? Further, Kaufman does not accept universal causal determinism, since he wishes to leave room for human freedom. This leads to other problems, however. If one accepts free will, one has accepted the reality of genuinely 'new' events, events not caused by their antecedents. But if we accept uncaused events in the realm of human action, why not allow for God to act in the world in a similar way, initiating new causal sequences that would not other-wise have occurred? Sometimes Kaufman seems ready to speak in this way, allowing for instance that 'certain of [God's] subacts are responsive to our acts'.[33] But what can this mean? Does God anticipate human free will and build his response into creation from the start (strong divine determinism), or does he have to *curb* human free will in order to make sure his predetermined plan comes to fruition? Neither represents a very attractive picture; the second may not even be coherent. Tracy concludes that a theology of divine action within a purely determinist framework, one in which God does not act directly in the world in some way, is inherently problematic.

If God cannot be the providential Lord of a fully determined history, what about viewing history as partially determined and partially open? Could there be 'particular providence without gaps'? In Tracy's typo-logy, Brian Hebblethwaite provides such a hybrid position. On this view, nature is 'flexible' enough to allow God to affect the outcome of world events *without* there being gaps in the causal fabric of the world. Hebblethwaite holds that God's pervasive influence can be efficacious without altering the (physical) causal structure in which they are embedded: 'The whole web of creaturely events is to be construed as pliable or flexible to the providential hand of God . . . [God makes] the creature make itself at each level of complexity without faking or forcing the story.'[34]

This answer is initially very attractive. Like Lewis Ford's 'lure of God', like process theological appeals to 'divine persuasion' and like Wolfhart Pannenberg's all-determining reality (*alles bestimmende Wirklichkeit*) which still leaves room for human freedom and responsibility, it seems to allow God gently to guide history in a redemptive direction, yet without dissolving the framework of natural law or heavy-handedly determining future outcomes. Yet, as they say, the devil lies in the details. If different outcomes are possible due to God's influence, then we *must* say that they are causally underdetermined, since the same antecedents admit of alternative results in principle. But this is just to say that determinism is false. Tracy may be wrong to jump immediately from this critique to the notion of 'gaps'. For theologians such as Pannenberg and Ford have developed a sophisticated understanding of causality which presents an open view of history taken as a whole, and Arthur Peacocke, to whom we return in a moment, gives a detailed account of 'top-down' causation. (That Tracy has not 'closed the gaps' against top-down accounts is clear from the fact that he lists three questions for Peacocke to answer in greater detail without, however, raising effective criticisms against the view.) Nonetheless, as long as one focuses on the individual event it is hard to see how the outcome wouldn't have to be *either* causally determined *or* open to variant outcomes.

Tracy also criticises the position of John Compton, who utilises the model of human action to understand divine action.[35] One can describe human behaviour either in terms of physiological processes or as a set of intentional actions by agents; for Compton these are 'coordinated but independent forms of discourse about human beings' (Tracy, p. 307). Compton thus concludes that, as long as God's relation to the world is analogous to our relation to our bodies, 'God does not need a "gap" in nature in order to act, any more than you or I need a similar interstice in our body chemistry' (p. 308). As Tracy realises, the main difficulty for this view is the relationship between the two ways of speaking. If one dichotomises between the language of physical causes and the language of God's intentions in history, then – however specific God's intentions might be – one can only really speak of God's *single act* in creating history as a whole, and one thereby falls back into Kaufman's waiting arms. (Tracy admits that this is actually the direction Compton leans.) If, on the other hand, one says that they are really just two different ways of speaking of the same process ('dual-aspect theory'), and that process is fundamentally physical, then Compton's approach is actually granting full physical determinism, and no event really involves a direct divine action within history.

There are approaches to the so-called mind/body problem for *human* agents that are much more hopeful than Tracy grants; we return to them in the next chapter. But he is surely right that there must be some place for divine action if it is really supposed to mean something in the world. As Tracy summarises, 'If by an "act of God in history" we mean a divine initiative (beyond creation and conservation) that affects the course of events in the world, then it is at least very difficult to see how such an action could leave a closed causal structure untouched' (p. 310). A full doctrine of God requires an open world, one with causal spaces in which God could act. Moreover, these must be such that natural law is not suspended or broken every time God acts, which would make a mockery of the natural order. Finally, it must be the sort of openness that will not be closed up by advances in scientific knowledge, leaving theologians stranded high and dry (again). The history of embarrassment is long enough already.

What kind of 'explanatory gaps' would do the trick? They are found 'when we are *unable* to give a complete account of the sufficient conditions for an event', or 'when our theories entail that human knowers will not [even] *in principle* be able to give a sufficient explanation of some of the events that fall within the theory's scope', or when we encounter events that 'are not uniquely determined by their antecedents'.[36] According to Tracy, gaps in principle are certainly depicted by chaos theory and *may* be illustrated by quantum theory as well. Of these two cases, up to now only quantum theory has revealed real causal gaps in nature's fabric, since it alone gives evidence of an irreducible indeterminism in nature. This fact is of potentially great significance:

> There is no competition with or displacement of finite causes here, since there is no sufficient finite cause that could explain why the probability function collapses as it does. The world at the quantum level is structured in such a way that God can continuously affect events without disturbing the immanent order of nature. (p. 318)

Note, however, that the openness alone is not in itself enough for theological purposes. Not only must the world have an open structure (i.e. not all outcomes being determined by their antecedents), and not only must any interruptions in causation be 'natural' and not 'disruptions', it must also be the case that underdetermined events are somehow amplified to the point that they can make a difference in the big picture.

Tracy adds that even if all these conditions were met, it might not yet be enough for a theory of divine action. For the sort of 'indeterministic chance' that we have been discussing is not yet *free intentional action*. Free choice is certainly more than mere non-determination; 'freedom requires *self*-determination; the agent must be able to decide which of the alternatives open to her will be realized in her action. So non-determination is a necessary but not sufficient condition for freedom, and advocates of this sort of freedom bear an additional burden of argument (i.e. beyond that born by someone who asserts simply the presence of structured chance in the world)' (p. 311). Here we find a fascinating interplay of scientific and theological parameters. A theory of divine or human agency can never be derived directly from the natural sciences (see Chapter 6 above). Yet the physical phenomenon of chance has significant theological implications, depending upon the precise nature of chance occurrences in the universe.

1. *Causal chance* involves the meeting of separate causal chains in a manner that is unforeseeable to finite agents (winning the lottery, meeting a friend in a foreign city).
2. *Chaotic unpredictability* makes the outcome unforeseeable in a much stronger sense: the outcome could not be known by finite agents *even in principle*. Here the 'veil of ignorance' cannot be rent: we *could* never know if God had intervened to adjust the initial conditions. Still, *given* the initial conditions, the outcome must follow; the equations of chaos-theory mathematics are deterministic.
3. *Only indeterministic chance*, then, provides the opening for divine action. Assuming that the limits of knowledge of the microscopic world (the Heisenberg uncertainty principle) are intrinsic, and assuming that they represent genuine (ontological) openness and not, say, the effect of hidden variables, real randomness is at work in nature.

If God is to use the world's openness providentially, as a means for bringing about his purposes, it must have two further features. First, the chance occurrences must still be part of an ordered system. As Tracy notes, indeterminacy is theologically useless

if [the chance] events simply represent random disruptions of otherwise orderly natural processes. Events of this sort could not be
given a coherent place in a scientific description of the world; the
world would not so much have an open structure as an incomplete
one. If we are to understand God to have an ongoing and pervasive
role in contributing to the direction of events, then the world must
be structured in a way that is both open and ordered, smoothly
integrating chance and law. (p. 316)

Second, openness at the quantum level needs some mechanism that
would amplify any divine guidance up to the macroscopic level. In
Tracy's words, 'Chance will be irrelevant to history if its effects, when
taken together in probabilistic patterns, disappear altogether into wider
deterministic regularities' (p. 317). The trouble is that statistical probabilities at the quantum level are usually said to cancel out at the macrophysical level. Given a large enough scale and low enough temperatures,
quantum effects do not prevent us, for example, from calculating plan-
etary motions using Newton's laws. (More significant are the limitations
expressed by special and general relativity.) If God acts occasionally but
stays within the statistical probabilities, the summation of the probabilities
will cancel out the effects of that action. If, however, God acts so regu-
larly as to radically alter the probabilities, then 'we are back to the
"interventionist" disruptions of the natural order' (p. 316).

It is important to acknowledge the limitations of this model of divine
action. Still, it continues to offer theology resources found in no other
area of the physical sciences. Given what we currently know, no laws of
nature would be broken even if God continually guided the resolution
of quantum probability states, amplifying the results through deterministic structures such as chaos into measurable changes in the physical
world (much as a Geiger counter amplifies radioactive decay into audible
clicks). As Tracy notes, guidance at the quantum level could produce an
otherwise unlikely brain state (which the individual might experience
as a particular thought) or a gene mutation that would have specific
effects. In principle, this could be done while preserving the expected
quantum probability distributions; only at the level of history as a
whole – and only for those willing to think in terms of a possible 'intention' and 'purpose' behind history – would the pattern be (potentially)
visible.

Tracy summarises five types of divine agency (p. 319), which might
serve for us as the five loci of the theory of divine action defended in
this book. In addition to the initial creation:

214

1. 'God acts directly in every event to sustain the existence of each entity that has a part in it' (the doctrine of conservation).
2. 'God can act directly to determine various events which occur by chance on the finite level' (quantum-level intervention).
3. 'God acts indirectly through causal chains that extend from God's initiating direct actions' (the amplification effect).
4. 'God acts indirectly in and through the free acts of persons whose choices have been shaped by the rest of God's activity in the world' (divine persuasion?).
5. 'God can also act directly to bring about events that exceed the natural powers of creatures, events which not only are undetermined on the finite level, but which also fall outside the prevailing patterns and regular structures of the natural order' (miracles in the classical sense).

After careful analysis and attention to the scientific details, we come to a surprising result: all but the last of these five types of divine action can be accepted without affront to natural law. Theologians will divide on the fifth, since many of us will find it difficult to claim to know that God has acted against the 'patterns and structures' of his own creation. Still, setting aside that area of disagreement for now, one cannot help but be encouraged by the openings for divine agency that we have found.

Strengthening the Theological Requirements

We have already encountered the work of Nancey Murphy above. In *Beyond Liberalism and Fundamentalism* she attempts to find a *via media* between 'interventionist' theories of divine action and the liberal holistic models that leave no place for specific divine action.[37] Through the work of the Vatican Observatory working group, Murphy has contributed to the theories put forward by Tracy, Wildman and Richardson, and others; her own account thus overlaps with theirs on a number of points. What is different about her emphases?

First, Murphy begins with theological criteria rather than describing the scientific results and then developing a theology of divine action to be consistent with them. As she has argued elsewhere, she presupposes that science and theology are epistemically on a par and that both must submit to criticism and correction from the other.[38] It is important to see how different this approach is from most of the authors so far considered. For Murphy, there is no greater initial justification in scientific explanations than in theological ones; in the case of conflict, either may have to give. As a result, she approaches the divine action question with a robust list of theological requirements[39] and then, given the scientific constraints, tries to make certain common metaphysical assumptions give way. Concretely, this means that theologians should refer to 'traditional formulations' of doctrine when evaluating the implications of science: 'Only if the formulations of the past turn out to be hopelessly unintelligible should they be rejected or radically changed' (p. 330). She also stresses two necessary conditions for a theory of divine action that is supportive of Christian theory: it must preserve 'special divine acts' which allow us 'to be able to distinguish between God's acts and the actions of sinful creatures' (p. 330). Additionally, it must allow for 'extraordinary divine acts', which Murphy seems to treat in the same way as traditional miracles except that they may not involve the notion of 'a violation of the laws of nature' (p. 331).

I am not sure that we need to take miracles or 'extraordinary divine acts', with or without law-breaking, as a theological requirement from the outset. William Alston has argued convincingly that in talking of 'special acts' of God, the specialness may refer to their effect on us rather than to the mode through which God brings them about: 'Just by virtue of creating and sustaining the natural order God is in as active contact with his creatures as one could wish . . . If God speaks to me, or guides me, or enlightens me by the use of natural causes, he is as surely in active contact with me as if he had produced the relevant effects by a direct fiat.'[40] What is emerging here – and presumably this is what Murphy intends as well[41] – is a new understanding of God's agency that is a *tertium quid* between, on the one hand, the traditional notion of a divine action as God's directly bringing about some result in the world and, on the other, the standard (post-Newtonian) model of natural processes which *indirectly* actualise divine intentions without any further activity on God's part. But it is not clear to me that continuing talk of 'extraordinary divine acts' is the right way to emphasise the *tertium quid* nature of this new alternative.

At any rate, with the traditional requirements for a theology of divine action in place, and with Murphy's commitment to taking scientific results seriously, clearly something else will have to give. As she writes, 'nothing short of a revision of current metaphysical notions regarding the nature of matter and causation is likely to solve the problem of divine action' (p. 334). (Disregarding current assumptions about matter and causation is justified because they are presently 'in great disarray'(p. 338).) Murphy claims that the current theories of God's causal agency (occasionalism, deism and double-action theory) are inadequate; we are in need of an entirely new causal theory. The new theory will preserve talk of lawlike regularities in nature. But in order to make a place for the new physics, for human free will and for divine causality, it will also challenge crucial features of Newtonian physics. Explicitly she challenges determinism, reductionist forms of explanation, and 'ontologizing' natural laws.[42] Implicitly the argument seems to involve a sustained questioning of *materialism* as a metaphysical assumption in the natural sciences. She discusses new trends that seem to suggest that 'matter is inherently active' (p. 336) and, although she questions 'top-down causation alone', she does seem to defend 'top-down causation by God', albeit 'mediated by specific changes in the affected entities' (p. 339).

Admirable in Murphy's treatment is her commitment to spelling out the *way* that divine, top-down causation would function. The answer is that God could guide the world at the quantum level without breaking natural law (miracles in the old, bad sense). Moreover, God could amplify these directing influences through chaos effects so that they resulted in significant effects in the world. The argument turns on the indeterminacy of quantum events. Murphy offers an analogy:

> Buridan is supposed to have hypothesized that if a starving donkey were placed midway between two equal piles of hay it would starve to death for want of sufficient reason to choose one pile rather than the other. I am supposing that entities at the quantum level are miniature 'Buridian' asses. The asses have the 'power' to do one thing rather than another... The question is what induces them to take one course of action rather than the other... (p. 341)

In the quantum case, God is the force that propels indeterminate states into one resolution or the other.

Another way to put the argument is in terms of a fourfold forced dilemma: quantum behaviour is either random, or internally determined, or externally determined (by some physical entity or system), or determined by God. Most physicists agree that the middle two options are unlikely. Genuine randomness is 'difficult for the scientific community to accept' because of the 'philosophical assumption that all events must have a sufficient reason' (p. 341). The best answer, then, is 'divine determination' of quantum events. Murphy concludes, 'To put it crudely, God is the hidden variable' (p. 342).

Ultimately, the metaphysical framework being offered here seems to represent, as Murphy writes at one point, 'a return to the Aristotelian view', which held 'that the form (organization, functional capacities) of an entity is equally constitutive of reality as is the stuff of which a thing is made' (p. 338). The view is also non-materialist and Aristotelian in allowing for the role of God in (apparently) all quantum events (note 33). Along with the physical influences on any quantum event there is also 'an intentional act of God to actualize one of the possibilities inherent in' it (p. 343). Theologically, this amounts to viewing all created entities 'as having "natural rights," which God respects in his governance' (p. 342). Another theological motif is the role of order in creation:

> Now, if the behavior of macro-level entities is dependent upon God's sustaining their specific characteristics by means of countless free and intentional acts, why do natural processes look so much like the effect of blind and wholly determinate forces? Since we have undermined the standard modern answer – determination by the laws of nature – a different account must be provided. The account to be given here is theological: one of God's chief purposes is (must have been) to produce a true cosmos – an orderly system. If we ask *why* God purposed an orderly universe we might speculate that it is for the intrinsic beauty and interest of such a cosmos. We could ground this speculation in our own aesthetic appreciation and in the supposition that our appreciation is an aspect of the *imago Dei*. (p. 346)

The significance of chaotic behaviour 'is that it gives God a great deal of "room" in which to effect specific outcomes without destroying our ability to believe in the natural causal order' (p. 348).

This is a very robust theory of divine action. Special acts of God occur constantly, perhaps in *every* quantum-level event (and recall that everything in the macrophysical world is built up out of quantum events). These acts of God direct the entire universe according to God's purposes; they may also impact on what we as humans think via a bottom–up manipulation of neural states. Murphy is inclined to say that 'God affects human consciousness by stimulation of neurons', causing particular thoughts to come to mind and presumably also producing new thoughts, a mechanism that would be sufficient to constitute divine revelation to individuals (pp. 349f.). The model also leaves room for knowledge of God's intentions in human history (via God's intentional actions) and for petitionary prayer (p. 352). She worries that at the level of observed phenomena the results would be consistent with deism and naturalism;

yet, she sees, this is in a sense a strength, since it keeps God's action hidden and ambiguous, leaving a role for faith.

Sometimes one does not know exactly what one is looking for until one sees the best available option and reflects on its implications. I suggest that this is what has happened in the case of Murphy's position. It remains theologically important that theologians can find *some* avenue by means of which God could exercise causal agency on the world, lest one's position on divine action be a de facto deism. Murphy has sought to show that both top-down influence and bottom-up causality via quantum-level phenomena are thinkable, and she has speculated that chaos effects might help make quantum effects felt far and wide. And yet one has the impression that the resulting quasi-Aristotelian view of non-material (divine) causes playing a pervasive role in the quantum world would call forth some raised eyebrows from most natural scientists. If the question is, 'Is it possible?' I think Murphy has made her case. But if one asks, 'Is it plausible?' many scientists, confronted with Buridan's-ass-like quantum particles and God as the hidden variable, may well have a different response. Perhaps it would be better, once the initial possibility of divine action has been won, to leave behind detailed questions of means and mechanisms. Otherwise the danger arises that our theories will sound Aristotelian in the sense that the early modern thinkers derided: making postulations about the nature of events in the physical world for which we have no empirical evidence.

Interestingly, Murphy seems to come to a similar recognition at the end of her piece, recognising that her position may appear like a 'two-language solution' (p. 354) similar to Kant's advocacy of the complementary languages of determinism and free will, each true in its own domain. As she correctly points out, an account like hers removes the contradiction, and thus the incompatibility often alleged between scientific and theological accounts of natural events. This is an important result. But what seems to occur, once it has been achieved, is that the scientist continues to pursue the best empirical, predictive and law-based knowledge she can achieve of the physical phenomena in question, whereas the theologian then finds herself freed to reflect further about God's nature and purposes in the universe based on a wider range of data, including scripture and human experience in the broadest sense. In particular, what needs to be explored, once Muphy's possibility of divine action is acknowledged, is the nature of the analogy between divine and human actors in the world – an analogy always in the background of Muphy's own argument as well. We turn to this task in the next chapter.

A Panentheistic Theory of Divine Action

No contemporary theologian has come closer to the panentheistic theory of divine action defended in this book than Arthur Peacocke; and of those who hold similar positions, none possesses a superior knowledge of the scientific developments. Peacocke's starting point, like Murphy's, is solidly theological:

> [God] is the 'Ground of Being' of the world; or for theists, that without which we could neither make sense of the world having existence at all nor of its having that kind of intellectually coherent and explorable existence which science continuously unveils.[43]

But – like Philip Hefner's well-known theory of humanity as God's 'created co-creator'[44] – Peacocke stresses that there is an ongoing creativity within creation itself: 'So we have to see God's action as being in the processes themselves, as they are revealed by the physical and biological sciences, and this means we must stress more than ever before God's *immanence* in the world' (p. 139). The most adequate way to think of this immanence is to understand the emergence of new forms – primitive life, higher organisms and human self-consciousness – as a result of God's immanent creative action in the world:

> Thus, the inorganic, biological, and human worlds are not just the stage of God's action – they *are* in themselves a mode of God in action, a mode that has traditionally been associated with the designation 'Holy Spirit,' the creator Spirit. I think that to give due weight to the evolutionary character of God's creative action requires a much stronger emphasis on God's immanent presence in, with, and under the very process of the natural world from the 'hot big bang' to humanity... The basic affirmation here is that all-that-is, both nature and humanity, is in some sense *in* God, but that God is, profoundly and ultimately, 'more' than nature and humanity. (p. 139)

In this treatment, as in his earlier major publication from which the idea is drawn,[45] Peacocke employs the metaphor of creation as an act of composing and of the created order as a musical composition: Beethoven is 'in' the present-day performance of his Seventh Symphony – even at the same time that the director and musicians are responsible for their own creative interpretation – in a way analogous to the creative role that God continues to play in the creativity of nature and humanity.

How does Peacocke evaluate the various avenues for divine interaction so far discussed? His responses to divine action at the quantum level, to 'amplification' by means of chaotic systems, and to 'quantum chaos' are uniformly sceptical. Peacocke finds the quantum solution no different than saying that God breaks the laws of classical physics by performing a miracle. For in the sorts of proposals we have been looking at, God would have to 'make some micro-event, subsequently amplified, to be other than it would have been if left to itself to follow its own natural course, without the involvement of divine action'. The only difference, he thinks, is that this type of intervention 'would always be hidden from us', whereas interventions in a Newtonian universe could be subsequently discovered.[46] I do not think that this response fully recognises the *tertium quid* nature of, say, the views of Nancey Murphy summarised above. Surely quantum-level theories of divine action involve God's influencing outcomes in the physical world; this is what separates such theories from the Wiles/Kaufman approach, which no longer identifies any particular locus (or loci) of divine action in the world. But precisely a panentheist such as Peacocke should recognize the *theological* worries associated with God's breaking natural laws. If the world is encompassed by (or, to speak metaphorically, 'contained in') God, then its regularities are inner-divine regularities, something like the autonomic functioning of our own bodies. Setting aside such regularities for the sake of focal divine action would amount to a sort of contradiction within the divine being, or at least a contradiction *in actu*. By contrast, God's acting through genuine openings in the fabric of natural law would not raise this problem. Thus, if there is to be focal divine action, the quantum possibility represents a distinct advantage.

Nonetheless, Peacocke's concerns about the scientific details of such an account ought to give one pause; they serve as a good reminder that an opening in the science/theology dialogue is not a conclusive answer. For example, God could act in this way only if we assume 'total divine omniscience and prescience about all events' (p. 155); but we may have *theological* reasons to resist this view. On the scientific side, there is as yet no clear understanding of how, if at all, chaotic systems could amplify individual divine actions on the quantum level such that they made a perceptible difference in the physical world. Moreover, so-called quantum chaos is a highly controversial hypothesis. Peacocke cites Bob Russell's statement: 'Whether or not quantum systems do actually display additional statistical behavior beyond that represented by the wave function – that is, whether or not there is a quantum version of classical chaos (called "quantum chaos") – is still an open question.'[47] Finally, the physicist Joseph Ford speaks of the 'major battle' being fought over whether such phenomena exist at all; at present, he concludes, 'the evidence weighs heavily against quantum chaos', which would represent 'an earthquake in the foundations of physics'.[48]

How then are we to view a theology of divine action that appeals to quantum indeterminacy? Recall that it is the *only* area of the physical world presently known to us in which God could bring about effects that would not otherwise have occurred without breaking natural law. Peacocke has urged caution. Our best science tells us that particular quantum phenomena are unpredictable *in principle* (just as it tells us that near-infinite knowledge of initial conditions would be required to predict outcomes in chaotic systems). If one takes the quantum solution, one attributes a knowledge to God which, although not involving logical contradiction, goes beyond what a finite knower could *ever* know about the physical world; one also assumes that God can bring about physical effects at the level of an individual electron or photon in a way that is incomprehensible to physics. For some this will not seem troubling: if God can count the hairs on one's head, why couldn't God know and do things at the microscopic level that no human could? For others, such as Peacocke, the problems with the details of this particular account suggest that theology is better off remaining at the level of *general* statements about God's purposes in the world. The theologian may note 'the new scientific awareness of unpredictability, open-endedness and flexibility and of the inbuilt propensies of natural processes to have particular kinds of outcomes' (p. 157). But she should not go beyond these features of the world to try to specify the 'causal joint' through which God brings about his purposes in the world.

According to Peacocke, then, we can never locate a locus of divine action within the interstices of the world and then conceive of it being amplified to affect cosmic history. If God is to act providentially at all, the influence will move not from the part to the whole but from the whole to the part. This he calls, following Campbell and Sperry, 'top-down' causation or 'whole part constraint'.

In scientific contexts, Peacocke admits, what one actually encounters are examples of the influence of 'boundary conditions' on the behaviour of members of the system: 'The set of relationships between the constituent units in the complex whole is a *new* set of boundary conditions for those units'.[49] One example would be the effect on a group of animals of their total environment, an effect reflected in the DNA of those that survive and reproduce most effectively. Another example involves the effect of mental phenomena on brain states, to which we turn in the next chapter. There are also purely physical examples of patterned behaviour where the patterns clearly emerge from the system as a whole and are not found in the particles individually. Peacocke notes:

> The notion of causality, when applied to systems, has usually been assumed to describe 'botton-up' causation – that is, the effect on the properties and behavior of the system of the properties and behavior of its constituent units. However, an influence of the state of the system as a whole on the behavior of its component units – a constraint exercised by the whole on its parts – has to be recognized . . . For, to take the example of the Bénard phenomenon, beyond the critical point, individual molecules in a hexagonal 'cell,' over a wide range in the fluid, move with a common component of velocity in a coordinated way, having previously manifested only entirely random motions with respect to each other. In such instances, the changes at the micro-level, that of the constituent units, are what they are because of their incorporation into the system as a whole, which is exerting specific constraints on its units, making them behave otherwise than they would in isolation.[50]

Many of the well-known chaotic phenomena thus fit within this category. While such behaviours are not 'emergent' in the strong sense, they go far towards illustrating the important influence of 'higher' or emergent levels on lower-level phenomena.

It is my contention that Peacocke's shift to 'top-down' causation as the exclusive mode of divine activity in the world importantly transforms the debate about divine agency. Indeed, one might say that it meets the goals of Nancey Muphy's above-mentioned book, *Beyond Liberalism and Fundamentalism*, in a way that is more consistent than her own position on quantum-level causation by God described above. On the one hand, Peacocke has to give up on direct physical causation by God. It cannot occur in deterministic contexts, since this would involve law-breaking by God, nor, he thinks, does it occur at the quantum level. In one sense, he has to admit, 'The word "causation" is not really appropriate for describing such situations,' namely the effect on the parts 'of being in the interacting, cooperative network of that particular, whole system'.[51] On the other hand, there is still ample space for speaking of God's directing of the world as a whole, hence of divine providence in this sense. This is where panentheism plays its most important role in the theory of divine agency: God, who contains the universe-as-a-whole (though he is much more than it as well) can act directly on the universe as its constraint and overarching context.

> If God interacts with the 'world' at this supervenient level of total-
> ity, then he could be causatively effective in a 'top-down' manner
> without abrogating the laws and regularities ... that operate at the
> myriad sub-levels of existence that constitute that 'world'.[52]

This holistic influence could operate, then, as a constraining influence
on more limited systems within the universe, for instance (eventually)
on the biosphere of our planet.

There are of course questions about this approach. One – the question
of whether whole-part constraint should really be described as causality
at all – we have already noted. Another concerns the indirectness of the
divine agency here imagined: how significant could God's guidance of
the individual be if it comes to her highly mediated? Would Peacocke
have to speak of a top-down hierarchy proceeding from the universe-
as-a-whole, down through superstrings and galaxies to our individual
planet, and then through the history of biological evolution and countless billions of genetic mutations to one person existing today? Such an
interpretation is certainly suggested when he writes that 'the state of
the world as a whole (all-that-is)' is 'the field of the exercise of God's
influence' (p. 161). But such a model is strangely reminiscent of Aristotle's
God, the *nous noetikos*, who transmits motion to the outermost spheres
of the heavens (by their emulation of 'his' perfection), and from thence
down through multiple spheres to the earth and its inhabitants, even
without any awareness on their part of what 'he' has done – hardly a
robust theory of providence! If this will not do, could Peacocke then say
that God is 'top-downly' active *directly* from the universe-as-a-whole to
me? How could such a direct causal influence be specified without the
directness and concreteness that worries Peacocke in the quantum-level
theological theories discussed above?

Although he denies miracles (breaks in natural law), Peacocke tries to
speak of God's activity in the strongest terms available to him. God is
present not only to the totality but, as one who is 'in, with and under'
the universe (p. 176), he is also present to each individual entity as well
(p. 162). Due to the nature of the panentheistic hypothesis, the creative
processes that make up the natural order are likewise 'themselves the
immanent creative activity of God' (p. 163). Since God is imparting
information, we can preserve the doctrines of divine self-communication, revelation and prayer. Since top-down causation allows us to regard
events as manifesting God's overall intentions, general providence plays
a role (p. 162), and even special providence is maintained:

> *particular* events or clusters of events, whether natural, individual and
> personal, or social and historical, (a) can be specially and significantly

revelatory of the presence of God and of the nature of his purposes to human beings; and (b) can be intentionally and specifically brought about by the interaction of God with the world. (p. 182)

Still, in the final analysis Peacocke has to admit that his view, like those of Kaufman and Wiles discussed above, allows for God's action only 'on the world-as-a-whole' (p. 163). *Humans* may find more divine meaning in some (natural or historical) events than in others, but 'God is *equally* and totally present to *all* times and places' (p. 181, my emphasis). When one reads that God can influence 'any constituent entity or event in the world that God wishes to influence' (p. 164), or that 'Particular events could occur in the world and be what they are because God intends them to be so' (p. 159), one must therefore interpret the influence as exclusively a holistic, totality-affecting one: God may be intimately present to each entity, but God acts *causally* on us all only from the (very) top down.

This limitation is perhaps moderately troubling; it becomes more disturbing when one turns to that central Christian concern, the revelation in the person and work of Christ. Consistent with the view just sketched, Peacocke attempts to use recent information theory in order to conceive of a 'naturalistic' model of revelation. In opposition to the common separation between revealed knowledge and natural knowledge he proceeds 'by presenting christology along lines that regard the self-communication of God to humanity as an "informing" process'.[53] The New Testament represents a 'development' of seeds of judgement and reflection on Jesus, rather than an 'evolution' with mutation. In short, we find in Jesus something like a set of 'God-driven' emergent properties, making Jesus a kind of ultimate emergent. Thus he writes, 'Might it not be possible for a human being so to reflect God, to be so wholly open to God, that God's presence was clearly unveiled to the rest of humanity in a new, emergent, and unexpected manner?'[54] But emergent properties emerge out of the world and are not imposed on it from outside: 'Taking the clue from the Johannine Prologue, we could say that the manifestation of God which Jesus' contemporaries encountered in Him must have been an emanation from within creation, from deep within those events and processes which led to His life, teaching, death, and Resurrection.'[55] In this way Peacocke explains that God did not enter an otherwise 'closed' world through Jesus, but only made himself known through Jesus, using causal mechanisms fully consistent with the rest of cosmic history.

Now Peacocke wants nonetheless to maintain that Jesus embodied and enacted some highly specific divine intentions.[56] Thus, for instance,

God 'communicated' through Jesus 'an explicit revelation of the significance of personhood in the divine purposes – an insight only partially
and incompletely discernible from our reflections on natural being and
becoming'.[57] Here he must mean more than that Jesus, being the kind
of person he was, was lucky enough to come to symbolise God's nature,
without God's intending that he do so, since in that case talk of 'revelation' would become equivocal. Rather, Peacocke writes as if Jesus'
resurrection were a deliberate act of revelation: the resurrection was an
event in which 'God was able to reveal further the way ahead for Jesus,
as "Jesus the Christ," to draw all humanity after him into a full relation
with God' (p. 308). The early witnesses, at least, interpreted Jesus and his
fate as 'a communication *from* God, a revelation of God's meanings for
humanity', and not just a humanly-authored parable of God's nature and
intentions (p. 296).

Yet without bottom–up causation God's role in the Jesus event can
be no more than the limited top–down influence we have already
explored. The degree to which God's influence on Jesus was mediated
becomes clear when Peacocke cites with approval a key passage from
John Bowker:

> It is credibly and conceptually possible to regard Jesus as a wholly
> God-informed person, who retrieved the theistic inputs coded in
> the chemistry and electricity of brain-processes for the scan of every
> situation, and for every utterance, verbal and non-verbal . . . [T]he
> result would have been the incarnating (the embodying) of God in
> the only way in which it could possibly have occurred . . . [W]hat
> seems to have shifted Jesus into a different degree of significance
> . . . was the stability and the consistency with which his own life-
> construction was God–informed.[58]

This position seems to bracket any particular intentions on the part of
God, viewing Jesus as someone who happened to activate a spiritual
potentiality already built into the created order – into his own genetic
code! – through a very long evolutionary process which the eyes of
faith can see as having been guided by God.

Many aspects of Peacocke's theory of divine agency are, *en fin de
compte*, theologically attractive. Since God's direction involves an 'input
of information' rather than energy transfers, the old dilemma of how
God as non-physical being could bring about physical effects disappears.
More importantly, Peacocke's theory of divine action is persuasive rather
than coercive. There is no greater respect for free decisions (and pre-
existing order) than in cases where an agent presents information and

allows it to be received and appropriated, or discarded, by other agents – exactly the model he advocates. Theologically, it is appropriate to speak of God's action holistically, at the level of all that is, while still emphasising God's presence to every individual. Finally, this model avoids that charge that God's actions are intermittent or sporadic; it finds divine providence at the most overarching level while still leaving room for special providence in at least one sense.

Nonetheless, I suggest that we will ultimately need to supplement the top-down theory of divine agency by something more. If the worries are justified that speaking of God's 'one act' is too little for Christian theism, as Thomas Tracy has argued, then we have reason also to worry about a God whose agency is top-down only. Given science, 'top-down' may well be the best way to conceive general providence, God's overall guidance of the world. But preserving the doctrine of special providence seems to require something more. The theologians working on a theory of quantum-level action (with or without chaos effects) have uncovered a remarkable possibility: that God might act in the world to guide events without breaking natural law or physical regularities. To take their work as settling the matter would be a mistake. But it does represent a possible locus of divine action unparalleled since Newton 'laid bare the inner workings of creation'. Combining the theories of top-down and bottom-up causation explored in this chapter provides, I think, a viable means of conceiving God's action in the world, one that is acceptable both theologically and scientifically. To complete the theory of divine action, we will need to think more carefully about the fundamental analogy between human and divine action – the task to which the final chapter is devoted.

NOTES

1. See Philip Clayton, 'The God of history and the presence of the future', *The Journal of Religion* 65 (1985): 98–108, and 'Being and one theologian', *The Thomist* 50 (1988): 645–71.

2. See Austin Farrer, *Faith and Speculation* (New York: New York University Press, 1967). Cf. also David Burrell, *Knowing the Unknowable God* (Notre Dame, IN: University of Notre Dame Press, 1986), as well as Burrell, 'Divine practical knowing: How an eternal God acts in time', in Brian Hebblethwaite and Edward Henderson (eds), *Divine Action: Studies Inspired by the Philosophical Theology of Austin Farrer* (Edinburgh: T. & T. Clark, 1990), pp. 93–102.

3. Thus my critique of Pannenberg's future ontology as 'counterintuitive' in 'Anticipation and theological method', in Carl Braaten and Philip Clayton

(eds), *The Theology of Wolfhart Pannenberg: Twelve American Critiques* (Minneapolis: Augsburg, 1988), must be taken as over-hasty. I offer a fuller and more nuanced critique in the article, 'Being and one theologian', cited above.

4. I here use the term in a way that is neutral on the question of whether these influences require the breaking of natural law.

5. Quoted in Mark Richardson and Wesley Wildman (eds), *Religion and Science: History, Method, Dialogue* (New York: Routledge, 1996), p. 239.

6. See Carl von Weizsäcker, *The World View of Physics*, trans. Marjorie Grene (Chicago: University of Chicago Press, 1952).

7. See John Polkinghorne, 'The metaphysics of divine action', in Robert J. Russell, Nancey Murphy and Arthur Peacocke (eds), *Chaos and Complexity: Scientific Perspectives on Divine Action* (Vatican City State: Vatican Observatory Publications, 1995), pp. 147ff., quote p. 155.

8. Maurice Wiles, *God's Action in the World* (London: SCM Press, 1986).

9. Nancey Murphy, *Beyond Liberalism and Fundamentalism: How Modern and Postmodern Philosophy Set the Theological Agenda* (Valley Forge, PA: Trinity Press International, 1996), pp. 71ff.

10. See Thomas F. Tracy (ed.), *The God Who Acts: Philosophical and Theological Explorations* (University Park, PA: Pennsylvania State University Press, 1994). Parenthetical references in the text are to this work.

11. See Kathryn Tanner, *God and Creation in Christian Theology: Tyranny or Empowerment?* (Oxford: Basil Blackwell, 1988), p. 89, also cited in Thomas F. Tracy, 'Divine action, created causes, and human freedom', in Tracy (ed.), *The God Who Acts*, p. 86.

12. See for example, Wolfhart Pannenberg 'Der Gott der Geschichte', in *Grundfragen systematischer Theologie: Gesammelte Aufsätze*, Vol. 2 (Göttingen: Vandenhoeck und Ruprecht, 1980), pp. 112–28.

13. I focus here on Polkinghorne's 'The metaphysics of divine action' in Russell, Murphy and Peacocke (eds), *Chaos and Complexity*, pp. 147–56.

14. David Bohm, *Wholeness and the Implicate Order* (London: Routledge & Kegan Paul, 1980); Bohm and B. J. Hiley, *The Undivided Universe: An Ontological Interpretation of Quantum Theory* (London: Routledge, 1993).

15. Austin Farrer, *Faith and Speculation* (New York: New York University Press, 1967).

16. See, for example, Hebblethwaite and Henderson (eds), *Divine Action*, especially the editors' Introduction. Especially good accounts of double agency can be found in this collection in Rodger Forsman, '"Double agency" and identifying reference to God' (pp. 123–42) and Thomas F. Tracy, 'Narrative theology and the acts of God' (pp. 173–96).

17. Farrer, *Faith and Speculation*, p. 154.

18. John Polkinghorne, *One World: The Interaction of Science and Theology* (London: SPCK, 1986), p.71ff.

19. This way of defending divine action in the world was classically presented by W. G. Pollard, *Chance and Providence: God's Action in a World Governed by Scientific Law* (New York: Scribner, 1958).

20. Polkinghorne, 'The metaphysics of divine action', pp. 152f.

21. Polkinghorne, *One World*, p. 72.

22. Numerous popular presentations of chaos theory are available, and certain apparent implications of the physics of chaotic systems, such as the so-called 'butterfly effect' (the motion of a butterfly's wings in China could in principle affect the weather in Los Angeles) have caught the public imagination. See, among others, James Gleick, *Chaos: Making a New Science* (New York: Viking, 1987).

23. See James Crutchfield, J. D. Farmer, N. H. Packard and R. S. Shaw, 'Chaos', in Russell, Murphy and Peacocke (eds), *Chaos and Complexity*, pp. 35–48.

24. Polkinghorne, *One World*, p. 74.

25. Wolfhart Pannenberg, *Systematic Theology*, Vol. 2, trans. Geoffrey Bromiley (Grand Rapids, MI: Eerdmans, 1994), especially pp. 44–6. Something like this view is also presupposed, I think, in the recent book by George Ellis and Nancey Murphy, *On the Moral Nature of the Universe: Theology, Cosmology, and Ethics* (Minneapolis: Fortress Press, 1996).

26. Richardson and Wildman, (eds), *Religion and Science*, p. 243.

27. Robert J. Russell and Wesley Wildman, 'Chaos: A mathematical introduction with philosophical reflection', in Russell, Murphy and Peacocke (eds), *Chaos and Complexity*, pp. 49–74.

28. Thomas Tracy (ed.), *The God Who Acts: Philosophical and Theological Explorations* (Pennsylvania: Pennsylvania University Press, 1994).

29. Thomas Tracy, 'Particular providence and the God of the gaps', in Russell, Murphy and Peacocke (eds), *Chaos and Complexity*, pp. 289–324. Parenthetical references in the text are to this essay.

30. Tracy's primary reference is to Schleiermacher's *The Christian Faith*, ed. H. R. Macintosh and J. S. Stewart (New York: Harper & Row, 1963).

31. See Wiles' essay in Owen Thomas (ed.), *God's Activity in the World: The Contemporary Problem* (Chico, CA: Scholars Press, 1983), already discussed above; Wiles' monograph, *God's Action in the World* (London: SCM Press, 1986); and Schubert Ogden, *The Reality of God and Other Essays* (New York: Harper & Row, 1966).

32. Tracy draws primarily from Gordon Kaufman's chapter, 'On the Meaning of "Act of God",' Chapter 6 of *God the Problem* (Cambridge, MA: Harvard University Press, 1972).

33. Kaufman, *God the Problem*, pp. 146f.

34. Brian Hebblethwaite, 'Providence and divine action', *Religious Studies* 14/2 (1978): 223–35, quoted in Tracy, p. 305.

35. See especially John Compton, 'Science and God's action in nature', in Ian Barbour (ed.), *Earth Might Be Fair: Reflections on Ethics, Religion, and Ecology* (Englewood Cliffs, NJ: Prentice-Hall, 1972).

36. Tracy, 'Particular providence', p. 290, emphasis added.

37. See Murphy, *Beyond Liberalism and Fundamentalism*. I recommend Murphy's book for a fuller picture of the historical and theological background to the contemporary debate as we have been examining it.

38. See Murphy, 'Postmodern apologetics, or why theologians *must* pay attention to science' and 'On the nature of theology', in Richardson and Wildman (eds), *Religion and Science*, pp. 105–20 and pp. 151–9.

39. See Nancey Murphy, 'Divine action in the natural order: Buridan's ass and Schrödinger's cat', in Russell et al. (eds), *Chaos and Complexity*, pp. 325–57. A theory of divine action 'must make sense of the traditional claim that God not only sustains all things, but also cooperates with and governs all created entities. This account needs to be consistent with other church teachings; it needs to leave room for special divine acts for both doctrinal and practical reasons; and it must not exacerbate the problem of evil' (p. 338). The following parenthetical references in the text are to this essay.

40. See William P. Alston, 'God's action in the world', in Ernan McMullin (ed.), *Evolution and Creation* (Notre Dame, MI: University of Notre Dame Press, 1985), pp. 213f.

41. As becomes even clearer in the final chapter of Murphy's recent book, *Beyond Liberalism and Fundamentalism*, cited above.

42. For a strong statement of the reasons to rethink the 'existence' of the laws of nature, see William Stoeger, 'Contemporary physics and the ontological status of the laws of nature', in Russell, Nancey Murphy and C. J. Isham (eds), *Quantum Cosmology and the Laws of Nature: Scientific Perspectives on Divine Action* (Vatican City State: Vatican Observatory Publications, 1993).

43. See Arthur Peacocke, 'Chance and law in irreversible thermodynamics, theoretical biology, and theology', in Russell, Murphy and Peacocke (eds), *Chaos and Complexity*, pp. 123ff., quote p. 138. Unless otherwise noted, all parenthetical references are to one of Peacocke's two articles in this collection.

44. See especially Philip Hefner, *The Human Factor: Evolution, Culture, and Religion* (Minneapolis: Fortress Press, 1993).

45. Arthur Peacocke, *Theology for a Scientific Age: Being and Becoming – Natural, Divine, and Human*, enlarged edition (Minneapolis: Fortress, 1993). For a thoroughgoing exposition of the views of Peacocke and Polkinghorne (as well as William Pollard) on divine agency, see Steven D. Crain, *Divine Action and Indeterminism: On Models of Divine Agency that Exploit the New Physics*, PhD Dissertation, University of Notre Dame, Notre Dame, IN, 1993, and his review article in *Zygon* 32 (Sept. 1997).

46. Peacocke, *Theology for a Scientific Age*, p. 154. Unless otherwise noted, subsequent references in the text are to this work.

47. Quoted in Arthur Peacocke, 'God's interaction with the world: The implications of deterministic "chaos" and of interconnected and interdependent complexity', in Russell, Murphy and Peacocke (eds), *Chaos and Complexity*, p. 268, note 11.

48. See Joseph Ford, 'What is chaos, that we should be mindful of it?' in Paul Davies (ed.), *The New Physics* (Cambridge: Cambridge University Press, 1989), pp. 366, 370, quoted in Peacocke, 'God's interaction with the world', p. 272.

49. See Peacocke, 'God's interaction with the world', p. 273.

50. Peacocke, 'God's interaction with the world', pp. 272f.

51. Peacocke, 'God's interaction with the world', p. 272, note 22.

52. Peacocke, *Theology for a Scientific Age*, p. 159. Again, the unmarked references that follow are to this work.

53. See Peacocke, 'The incarnation of the informing self-expressive word of God', in Richardson and Wildman (eds), *Religion and Science*, p. 321.

54. Peacocke, 'The incarnation', p. 332.

55. Peacocke, 'The incarnation', p. 331.

56. I am grateful to Steven Knapp for detailed correspondence on this topic and owe to him some of the formulations that follow.

57. Peacocke, *Theology for a Scientific Age*, p. 305. The following references are to this work.

58. John Bowker, *The Religious Imagination and the Sense of God* (Oxford: Clarendon, 1978), pp. 187f., cited in Peacocke, *Theology for a Scientific Age*, p. 298. See also Bowker's earlier *The Sense of God* (Oxford: Clarendon, 1973).

8

UNDERSTANDING HUMAN AND
DIVINE AGENCY

We need to cultivate a vision of reality (a metaphysics) that makes it [reality] truly independent of our given cognitive powers, a conception that includes these powers as a proper part. It is just that, in the case of the mind-body problem, the bit of reality that systematically eludes our cognitive grasp is an aspect of our own nature. Indeed, it is an aspect that makes it possible for us to have minds at all and to think about how they are related to our bodies. This particular transcendent tract of reality happens to lie within our own heads. A deep fact about our own nature as a form of embodied consciousness is thus necessarily hidden from us.[1]

In previous chapters we have examined the various types of activity in the world that Christians have traditionally attributed to God. A variety of theologians have appealed to recent developments in physics as offering ways to think of divine action in the world in a manner consistent with what we know through science. Each of the authors has attempted to avoid 'God of the gaps' strategies. A God who is assumed to be active only where natural science cannot yet answer some particular question about physical causality is a God who will be removed from the stage as soon as scientists close this particular hole in their account of the world.

We also entered into the difficult but essential debate over how the divine activity is to be thought about given what humanity now knows of the physical world. The reader is now familiar with the major accounts of divine agency developed in the recent literature: bottom-up, top-down, holistic, emergent, primary and secondary causality. Time and again, we found that the discussions turned on what it is to be a *human* agent in the world. Always at the pivotal point of his or her argument for divine agency, it seemed, the author would appeal to *human agency*, with this appeal often carrying the bulk of the argument's weight. Indeed, on reflection, this is perhaps as it should be: if one is to understand the agency of a personal divine being at all (and no one claims that we are going to understand it fully), then surely divine agency will have to be related *in some way* to the only other type of personal agency familiar to us: our own. Divine agency will be higher, of course, more complex, more powerful, more all-encompassing and more mysterious to us. But if it is nothing like human agency, we shall know nothing whatever about it.

Let me say it again: we have found that, despite many thinkers' intentions to rely on the physical sciences alone, the bothersome problem of *the human dimension*, the so-called problem of consciousness, emerged repeatedly as the linchpin of the entire debate. Indeed, the role of this particular problem in the discussion between the natural sciences and theology is so central that it suggests a new thesis: until one is able to solve the problem of human mental causation – the question of how human intentions and desires get translated into events in the physical world – one will not be able to develop even a half-way adequate answer to the question of divine causality. Conversely, *if* one is able to conceive of human intentional action in a way that is compatible with natural scientific accounts of the physical world, then one will have done the bulk of the work necessary for a theory of divine causation. For if the human spirit can produce events in the physical world, if our understanding of natural law allows for (or even demands) this type of causality, then divine action could be construed as an analogous mode of change in the physical world. The distinctiveness of divine action in contradistinction to human action must still be thought – but this is a task for which Christian theology has a great number of resources at its disposal.

My thesis, in short, is the following: the question of God's relation to the world, and hence the question of how to construe divine action, should be controlled by the best theories we have of the relationship of *our* minds to our bodies – and then corrected for by the ways in which God's relation to the universe must be *different* from the relation of our mental properties to our brains and bodies. The idea of such an analogy

233

is not new, of course. It motivated much of Charles Hartshorne's work on the God/world relationship earlier this century, and Schubert Ogden, among others, gave clear expression to the analogy several decades ago:

> I hold with Hartshorne that the interaction between God and the world must be understood analogously to this interaction between our own minds and bodies – with the difference that the former interaction takes place, not between God and a selected portion of his world (analogous to our own brain cells and central nervous system), but between God and the whole world of his creatures. Because his love or power of participation in the being of others is literally boundless, there are no gradations in intimacy of the creatures to him, and so there can be nothing in him corresponding to our nervous system or sense organs. The whole world is, as it were, his sense organ, and his interaction with every creature is unimaginably immediate and direct.[2]

I have called this argument the panentheistic analogy. It has sometimes been held that only non-classical theists can avail themselves of the analogy. But, as Charles Taliaferro notes in his excellent book on consciousness and the God/world relation, even 'classical theists can go so far as to maintain that the world is very much like God's body, even though this analogy must be very carefully hedged'.[3] The task for us in this final chapter, then, is to make progress toward a theology of the God/world relation by thinking our way more deeply into the so-called mind/body question, making full use of developments in the theory of human personhood (neo-emergentism, supervenience theory) in recent years.

It turns out that the two most important variables in this discussion are one's theory of the relation of the human 'mind' to its body and one's theory of God's relationship to the world. If humans have a purely spiritual soul associated with their body, one that can influence events in the physical world, then there is in principle no difficulty with a spiritual being such as God influencing events in the physical world. It is only necessary that God's relation to the world be understood as sufficiently analogous to our relation to our bodies for similar principles to apply. By contrast, what if human mental functioning can be fully explained in terms of neuro-physiological states and laws, and in this sense be reduced to them? In that case, it would appear, everything that 'folk psychology' calls 'mental' or 'consciousness' would be better accounted for in physical terms. It would seem to follow that there could be no real mental or spiritual causation within the world – or at least none that could be intersubjectively known. If physicalism (whether 'reductionist' or not) wins this battle, I will argue, the most hopeful means for making sense of divine causation will be removed. Perhaps faith can continue to assert divine action, but the grounds for our understanding will be reduced.

Hence it behoves us to look carefully at some of the main theories of mind on the market today, to compare their strengths and weaknesses, and to see whether the strongest among them offers any help in thinking through the knotty question of God's actions in the world.

PARAMETERS FOR A THEOLOGY OF GOD AND WORLD: THE PANENTHEISTIC ANALOGY

Problems with Dualism

It is important to note that the analogy between the soul/body relation and the God/world relation is as old as the theological tradition itself, and in fact even older. Plato, for example, held that the soul is what has the most reality, whereas the body (like all matter) is less important, less real, less good. Congruent with this position Plato argued that God is a spirit ('demiurge') who helped to organise pre-existent matter. Aristotle, who maintained that the soul is the principle or 'form' of the body, also held that the highest being, 'thought thinking itself' (*nous noetikos*), was the principle of motion and the final cause for all material and heavenly bodies. Thomas Aquinas also drew a clear parallel between the two principles: 'We find a certain imitation of God in man, . . . in that all man's soul is in all his body and again all of it in any part of the body, in the same sort of way as God is in the world'.[4] Descartes, who absolutely separated mind and body as two separate substances – indeed, to such an extent that any interaction between them became inconceivable – also separated God from the world so firmly that the problem of God's activity in the world became very difficult to conceptualise.

The dualism implicit in these earlier construals of the relationship between mind and body (and God and world) has been widely rejected in our day. If doing theology with an eye to the results of science means anything at all, it means working with theological proposals that do not presuppose a firm dualism – that is, the Greek dualism of two separate substances, 'mental' and 'physical' stuff. There are many ways to conceive of a closer relationship of mind and body – and thus a more holistic concept of the human person – without espousing a type of physicalism that would be incompatible with the tradition of Christian thought. Further, if we are successful in finding a mediating alternative, the panentheistic analogy will suggest that the relationship of God and world must *also* be thought of in a more holistic manner. Just as we do not seek a duality within the human person that leaves her fundamentally divided, but rather an integrated set of connections between her mental and physical functions, so also we seek a theology that allows for the full presence of God in, with and for the world created by him, without reducing God to the world or to a consequence of the world.

Before proceeding to the theory itself, it is important to note the convergence of factors that have led to the demise of strongly dualistic theories of the person. First, many of the motivations within philosophy for espousing dualism have disappeared. There was a time when philosophy and 'science' (i.e. natural philosophy) went hand in hand in supporting an ontology of multiple kinds of things or 'substances'. Since 'mind' and 'body' had radically different types of properties, they were assumed to be fundamentally different kinds of entities and were categorised as such. But the very ontology that kept philosophy in tune with the science of that day would set it at loggerheads with contemporary science. Therefore, philosophers have by and large rejected such views, with few today maintaining a dualism of mind and body. Much more common are views that either identify the two or show a dependence of mind on body in some way (see below). Likewise, recent exegetical and theological work has emphasised the holistic view of the person in the Old and New Testament and in the Christian tradition.[5] Pauline anthropology does not conceive of a state of disembodied existence as natural for the human being but emphatically understands the ideal state of existence after death as that of a mind embodied in a 'new heavenly body' (Phil. 3:21; 1 Cor. 15:35ff.). The Thomistic tradition has also argued vehemently that the idea of a soul disconnected from the body is philosophically difficult and theologically unacceptable.

The traditional view had held that, although the soul is located within the body in this life, it is essentially a separate 'thing' capable of existing apart from and outside of the body, at least for the period of time between the death of the body and its rising again as a new heavenly body.[6] As Jantzen notes:

> Such a picture of human personhood [that is, one that understands the soul as essentially transcending the body although it may dwell for a time within one] has been increasingly undermined as unable to cope with the combined protests of psychology, physiology

and philosophy, not to mention those coming from within theology itself; and today few people would be willing to accept a view which does not take the body seriously as essential for human wholeness.[7]

The Panentheistic Analogy and Its Limitation

The goal of a theology of the mind/body relation is to do justice both to the theological tradition and to the best science of our day. Repeatedly during the history of Christian thought, theologians have worked against the backdrop of a strong philosophical dualism; this was true of the Patristic era, of the period from Augustine to Anselm, and of the early modern period. As we have just seen, however, there are numerous reasons now to be sceptical of such dualistic separations. Our best knowledge of the physical universe and of the human person, and the dominant position within contemporary philosophy, is anti-dualistic.[8]

As it turns out, there are a number of reasons from within the sciences to resist strict dualism. Not least among these is the fact that science by its nature stresses what is observable or empirically testable. This stress led to the rejection of Aristotelian final causes at the dawn of the modern age, and it fuelled the scientific outrage at the theory of 'entelechies' or internal purposive principles at the end of the nineteenth century. It would be the best news for a knowledge endeavour such as science if 'mind' turned out to be fully reducible to observable and testable occurrences such as those studied by neurophysiology. At the very least, scientific method inclines scientists to speak of 'mental functioning' and 'mental predicates' rather than of a soul, which would have to contain ontologically unique mental states. Consequently, the view of humanity most congenial to science is that of a physical organism which evidences mental 'behaviors' – behaviours that may indeed require special sciences to study them (psychology, sociology, anthropology) but that are nothing above and beyond the body with which they are associated.

Note what happens to the theology of God and world if we apply the panentheistic analogy to this sort of account of the mind/body relation without supplementing it in any way. God becomes another word for the 'spiritual' – or perhaps mental – phenomena that occur within the world. This result may not bother the scientist: complex physical structures might evidence by-products that are not themselves purely physical even though they have physical causes. Such incidental qualities are often referred to as 'epiphenomena', which are appearances such as the Northern Lights that have physical causes though they may be perceived by us as 'something more'. God understood in this sense would be, it seems, purely immanent; but there would be no place for speaking of an actual transcendent being, one who might survive the Big Crunch and the end of the physical universe. Likewise, a God understood *only* through the scientific analogy could not *precede* the physical universe nor guide its development providentially. At best one would seem to have some form of pantheism. These worries will concern us in detail in the final section.

Panentheism with Transcendence

The sorts of conclusions just listed stand far from the tradition of Christian theology. Must theologians be driven to some such conclusions if they take science seriously? I do not think so. It is possible to pay very close attention to scientific conclusions and *at the same time* to recognise that the scientific results require an interpretive framework not dictated by the results themselves. Physics (and biology and neurophysiology) *underdetermines* its metaphysics; multiple metaphysical perspectives can 'interpret' the results – though some perspectives are more justified than others.

Note, for example, six metaphysical 'interpretations' that might be given of human mental experience. I list them in order of increasing boldness:

1. One could hold that there is a certain inevitability to the arising of mind or spirit (the anthropic principle).
2. It could be argued that mental functioning, once it arises, must remain an intrinsic part of this universe, and that in some way it will become the guiding force in the further evolution of the universe.[9]
3. One might maintain that mental predicates, once they have emerged, are manifestations of a 'thing' such as the soul that is really mental, a different kind of thing from the physical universe out of which it arises.
4. One could hold that it was inevitable that this new sort of thing, mind or spirit, would arise in the history of the universe. One might even hold that it is the fate of the universe to produce mind or spirit. Thus John Leslie has argued that objective value or good is built into the nature of the universe.[10]
5. One could argue that spirit has been present in some way from the very beginning, somehow preceding humanity and even life itself, perhaps helping to bring about the higher life forms at the

appropriate time. This would be a *teleological* view, since spirit would be exercising a purpose and bringing about a goal from the very beginning.

6. Finally, spirit could be taken as in some sense preceding the universe and transcending it, while at the same time being manifested within it. This view comes closest to classical theism, though it is also compatible with the panentheism defended in the present work.

We will not cover all the pros and cons of these six alternatives here. For now it is sufficient to emphasise that the decision between them is a *meta*physical decision, one not decidable on scientific grounds alone. The approach and mentality of science may perhaps incline one toward the earlier options rather than the more metaphysical responses at the end of the list. Still, there is no scientific reason to exclude the last options in principle, since scientific results alone do not seem capable of determining whether they are actually true or not.

As we have seen, the first battle for the theologian to fight is on behalf of the reality of mental phenomena, their irreducibility to the realm of the physical, and their real causal power to bring about either other mental phenomena or changes in the physical world. These are in the first place questions that arise within the disciplines that address the so-called mind/body problem: biology, physiology, cognitive science, psychology and philosophy. To the extent that she genuinely enters into the disciplines in question, the theologian can make real contributions to this discussion. *After* one has made a case for the reality of mental causes (as we will do below) – and thus for a model of the human being as both mental and physical – then the broader question of the analogy with the God/world relation arises. As we will see, the panentheistic analogy allows for some minimal, though still significant, theological inferences. But it will also need to be supplemented: the brain sciences have no reason to assert the continued existence of mental functioning once the body ceases to operate (corpses don't speak), whereas theologians have held that something about the human individual is destined to continue beyond the death of the body. Analogously, most theologians maintain that there is something in the divine nature that precedes the world, guides its evolution and continues in existence after its end – even if no physical science could give grounds for these assertions.

It can be a fine line that theology treads here: to refuse to supplement scientific results at all leads to pantheism – or at most to an emerging God with a finite lifespan; to assert traditional views with no regard for developments in neuroscience theatens to put theology and science into two incompatible realms. If God is understood as a being utterly separate from the world, for example, the problems with divine interventions *in* the world encountered above arise once again. Moreover, given an analogy between God and human subjects, human individuals would seem to consist of a spirit sharply distinct from the body it inhabits, giving rise to the famous Cartesian problem of how minds can influence bodies at all. Finally, under this view the human individual *in toto* becomes something that science can no longer study, since what we call the individual would become an uncomfortable union of two radically different components, one of which in fact transcends the world and is not a product of it. Only if God and the world are considered to be more closely linked than the tradition of 'utter transcendence' in theology allowed – only then can one develop a holistic view of the human person that preserves a place both for the scientific study of the individual and for the irreducible nature of the mental life.

Of course, conservative theologians for whom the panentheistic analogy 'goes too far' might be tempted to abandon the analogy between the divine and human subjects altogether. Certainly they have in Barth and his rejection of the *analogia entis* a clear precedent. However, the costs of rejecting the analogy are greater than is often acknowledged. As Jantzen writes in criticism of Barth, 'If divine transcendence is interpreted as complete otherness, total contrast with the world, then nothing revelatory could in principle occur *within* the natural order, since it is by definition utterly different from God and hence could not reveal him.'[11] Under this view, Jantzen points out, it is more difficult to account for interactions between God and the world; a greater distance between God and world is posited, such that it becomes hard to understand how it could be that 'the word became flesh and dwelt among us . . .' Colin Gunton has given forceful expression to the problem of postulating the total otherness of God. Supernatural reality then becomes

> precisely what nature is not, and therefore its relations with nature are necessarily problematic, just as Descartes found it impossible to reach a satisfactory understanding of the relation between mind and matter precisely because he had begun by defining them in opposition to one another. In particular, any historical activity of God will tend to take the form of an isolated intervention which is also a violation of the natural.[12]

With costs like these, we are better advised to move as far as we can with the analogy. If theological considerations later force us to modify it, we will be prepared to do so; until then we should follow it as far as it will take us.

THEORIES OF PERSONHOOD AND
THEOLOGICAL PREFERENCES

It is no coincidence that the view of the God/world relation in classical Western theism and its view of the human person have proceeded in parallel step. God was understood to transcend the physical world, to be essentially separate from it. For theologians such as Augustine, God's essence as timeless and eternal contrasted sharply with the pervasive temporality of the physical world. The divine nature was simple and unchanging, spiritual *as opposed to* physical. God did not need the world. After high Scholasticism, as the Aristotelian categories came under increasing criticism (say, as early as William of Occam), theologians argued that the world was to be understood according to its own laws, although the finite order as a whole still had to be sustained by God. Similarly, the human person consisted of an uncomfortable wedding of spirit and body. For the 'Platonic' theologians, human being depended on a dualistic (Cartesian) link between a soul-substance and a body-substance, between mind and matter. Many conceded to scripture that the natural state of the spirit was to be united with a body, but this view fell far short of a conceptually integrated view of the human person.

Under these conditions, divine action in the world could only represent an incursion in from another order. Any action that God accomplished would be in radical discontinuity with human actions in the world. Humans did not intervene in, much less interfere or conflict with, the world, whereas divine actions did all of these things. Note, however, that twentieth-century theologians have been less inclined to draw sharp discontinuities of this type. Karl Barth may still have understood divine action as a radical inbreaking into the natural world, but he still approached divine agency and divine subjectivity in a manner that showed the influence of the German Idealists and their stress on the structures of self-consciousness. According to Pannenberg, the Reformation theologians and their followers were wrong to interpret God's actions in light of timeless laws, as if God could 'timelessly resolve' to act in a certain way. Even Hegel comes in for criticism for trying to understand divine subjectivity in terms of the *necessary* laws of the development of self-consciousness. Instead, Pannenberg emphasises the radical divine freedom within the trinitarian circle of life, understanding each divine person as a subjective centre of activity, and thus as at least structurally similar to human subjects. The analogy between divine and human action is even more pronounced in the important recent work of Joseph Bracken.[13]

The analogy between divine and human subjectivity is also visible in the case of panentheists like Moltmann, McFague, Jantzen and myself, since we condone the analogy between the human mind's relation to its body and God's relation to the universe. It now seems that this analogy represents one significant argument in favour of panentheism. As we have come to see, doctrines of divine agency are not unproblematic for those who use the natural sciences as resources for explaining the natural world. It may be that contemporary science has discovered some significant 'spaces' through which God could influence the physical world (Chapter 7). But if there is *no* analogy between what we mean when we call God an agent and what we mean when we call humans agents, then it would seem impossible to develop a theology of divine action in any comprehensible sense. Indeed, the theology of divine action would seem to require there to be *some* resemblance between what it means to *act* in the two cases – some analogy between the way God influences events in the physical world and the way that humans actualise their intentions.

There is a further benefit to the analogy: if God's action is analogous to human action – that is, the word 'action' is not equivocal in the two cases – then divine action is not only acceptable (and at least partially comprehensible) but it is also less likely to be in conflict with natural laws. For clearly if it is a fact that human minds influence the physical world (as folk psychology, and each one of us in our everyday lives, presupposes), then our doing so cannot be in conflict with natural law! Instead – again assuming that human agents sometimes bring about events in the physical world – it must be a regular, non-contradictory feature of the physical world that humans have these effects. In short: natural law cannot be in conflict with human agency if agents act in the world as we think they do. Likewise, if God acts in a manner analogous to human agents, then it cannot contradict natural law that he does so. I thus conclude that panentheism, as the position that draws the clearest and most explicit analogy between human and divine action, also offers in principle the most powerful mechanism for making sense of divine action.

The First Criterion: The Irreducibility of Consciousness

What then is the best account of human action in the world? There is a variety of positions, as we will see in a moment. From a theological perspective, one issue rises in importance above all others: the debate over the reducibility or the irreducibility of consciousness. A recent international conference at the University of Arizona, for example, entitled 'Towards a New Science of the Mind', offered a clear instance of the dichotomy among some of the world's top specialists in this field. The one side held that consciousness will gradually become better and better grasped, and eventually fully understood, as we unravel the mechanisms of the brain. Nothing 'other-worldly', nothing qualitatively different from what the brain sciences can study, is involved in human thought, any more than scientists need to introduce some vitalist principle in order to explain the behaviours of cells.[14] The other side held that consciousness is forever beyond the conceptual tools of the natural sciences; neurophysiology may be able to describe *some* of the physical events that occur in conjunction with the mental life, but consciousness will never, even in principle, be fully explainable in terms of brain states and operations. There is an inner dimension of thinking, intending and willing, an inner sense of what it is like to be oneself and to think as one does, that is intrinsically irreducible to neurophysiological description.[15]

It is my view that Christian theology can and must side with the irreducibilists. One reason is that if minds were reduced by explanation to their physical substratum or 'hardware', then God would have to be reduced in a similar manner as well. (Conversely, if God does not need to be reduced in this manner, then it is unclear why human minds must be reduced.) Another reason is that the Christian eschatological hope of a life after death in a non-physical state would be impossible if the mind is no more than the physical states with which it is connected in this life.

Of course, knowing with whom one must side is a far cry from having made one's case successfully. The major task facing theologians in this area – and if I am right, the major task facing any theology of divine agency today – is to show why reductionist accounts of mind are inadequate on their own terms. This means addressing the position of 'eliminative materialism', which seeks to eliminate any explanatory role for mental predicates and instead to give an account of our cognitive life in consistently material terms.[16] But the theological task, I think, is also to challenge approaches that maintain that all existing things are physical things. Part of winning the case for non-physical explanations is showing why our explanations should include terms that refer to non-physical things.

The difficult question is how we are to provide a positive description of 'mental' phenomena that lets them be more than physical states and operations. Without question, classical Christian theology often allied itself to a metaphysical dualism all too reminiscent of Descartes: two kinds of things or substances exist, bodies and minds, and the human being is a compilation made up out of both sorts of 'stuff'. Of course, there are serious philosophers today who are dualists; Frank Jackson and Thomas Nagel, for example, both support a dualist ontology that takes conscious states to be more than physical phenomena.[17] Still, it seems to me that Christian thought has unnecessarily compromised its credibility because of its frequent marriage to this particular philosophical framework, the problems with which have been widely discussed in recent decades. (This marriage is ironic in view of the fact that the biblical texts, and particularly the Hebrew texts, do not presuppose an ontological dualism; it is a later addition which we have come to view, incorrectly, as indispensable.)

The Alternatives to Dualism

What other moves are left open to the 'irreducibilist' if a strict dualism of substances is left behind as unacceptable? Three alternative theories have been offered in recent years: the functionalist view, dual-aspect theory and supervenience theories. We examine the supervenience position in the next section. Functionalists take the mental as a series of functions or behaviours rather than as keys to reconstructing the mental life. Thus thinking, feeling and willing can be described only in terms of their external manifestations. Officially, functionalists merely decline to speculate on the metaphysical basis of these functions. Thus, for Ned Block the mind remains a 'black box' and we need be concerned only with the inputs (from the physical world) and the outputs (verbal behaviours, what one's body actually does – all verifiable events in the physical world).[18] Unofficially, however, functionalists tend to be highly sceptical about the existence of any non-physical entities or occurrences. I take functionalism to be not so much an answer to the mind/body debate as a refusal to address its metaphysical issues. On the other hand, its self-limitation to behaviours that can be empirically studied has made it a highly successful research programme in cognitive psychology.

Dual-aspect theory views the mental and the physical as two different aspects of the one world; thus it is just as accurate to speak of the one world in mental as in physical terms. The world itself is held to be neither – or to be simultaneously both. This view, which is much more at home in an Eastern, pantheistic framework and which is linked to the metaphysics of Spinoza, raises problems for Christian theism whether understood in classical theistic or in panentheistic terms. For all that it will say is that God is an aspect of the world – perhaps the sum total of mental states or conscious ideas, perhaps the world taken as a whole. By contrast, panentheists maintain that God both encompasses the world and possesses a mental life that transcends the world. Clearly this makes dual-aspect theory an opposing position and not an ally to the theologian (which is not to say that we have shown it to be false).

The Goal

So what *kind* of a position would the Christian theologian be inclined to espouse? What criteria would she bring to this debate, and what would she hope to attain? The position most congenial to theology would not be reductionistic; it would not make mind merely one way of viewing the totality of what is. It also would not reduce mind to being only a function that occurs, mysteriously, in an otherwise physical world, but would instead defend the reality of mental states in some sense.

Traditionally, theologians have held that these mental states are properties not of physical things such as brains alone, but of an active centre of mental activity – that which 'has' the mental states – which is ontologically distinct from the physical world. The tradition referred to this entity as the soul. I do not think it is necessary, or even possible, to start with a soul-substance in debate with the neurosciences today (though it remains to be seen how close we can get to it in the end). By this I mean it is not necessary to conceive of consciousness as a separate substance, which would dichotomise the world into two separate *types* of things.

Nonetheless, I think it is crucial for theologians to defend the irreducibility of consciousness, which is a type of *activity* that is essentially distinct from physical activity and causes. Consciousness requires a different set of predicates to describe it and a different type of explanation to make sense of it. But why this distinction; why reject soul-substance and retain consciousness as activity? The reason is that science and theology can make contact at the level of mental *activities*, whereas the introduction of an essentially non-physical thing such as the soul closes the door to such contact. Interestingly, the modern thinker to whom we owe this insight is Hegel. In the Introduction to his *Phenomenology of Spirit* he argued that substance must now be understood as subject, and subject as the activity of consciousness coming to itself. Only a theology of mental activities will be able to relate mind and body in a way adequate to the demands of contemporary cognitive science. And only such a view, it seems, will be able to preserve a place for the activity of God within a physical world understood so fully in physical terms. We turn now to a fuller explication of these views.

Review of the Argument against Materialism

In Chapter 6 we confronted the widespread argument that science leaves no place for the agency of God. The doctrine associated with this position is usually called *materialism* or *physicalism*. It is the view that all causes of events in the world are purely physical in nature – and therefore that all existing things are physical things, that is, things exhaustively describable in terms of physics. Although I challenged physicalism, we did discover that the explanatory success of the natural sciences casts doubt on claims to *know* that some individual physical event was caused directly by God. The result of that discussion was a presumption in favour of naturalistic explanations where they are available.

This negative result still leaves at least two non-physicalistic options available to the theologian or religious believer. First, it remains open to her to *judge* that God was active in bringing about a particular event. When she makes this judgement, she actually changes the genre of explanation, shifting from the framework of physical events and their (lawlike) causal histories to the framework of *actions*, which are behaviours carried out by agents with some purpose or intention in mind. To change genres in this way is to make a metaphysical move, turning from physicalism as the final word on the nature of the universe to theism, the belief that the universe is the product of an intentional agent who is in some way guiding its history according to a divine purpose and plan. The shift in explanatory genre makes clear that judgements about divine action are in the first place judgements about the overall nature of the universe, and only secondarily (i.e. by implication or inference) judgements about what caused some particular event in the world (say, your surviving your recent automobile accident).

A second option available to the theologian is to locate the general *limits or boundaries* of scientific explanations. When one has done this, one may find it possible to formulate theories about divine action that do not conflict with any (actual or, as far as one knows, possible) scientific accounts of the world. This task represented the central focus of the previous chapter: How precisely can one conceive of God acting in the world in a way that does not conflict with scientific accounts of physical causes? We looked in detail at recent important theories of divine action developed by scientist-theologians such as Polkinghorne and

Peacocke and philosophical theologians such as Wiles, Tracy and Farrer.[19] I agreed with several thinkers (e.g. Russell, Murphy and Tracy) that quantum indeterminacy meets the conditions for an understanding of divine agency that does not clash with any (actual or possible) scientific causal account and thus is not 'interventionist' in the perjorative sense of the word. It is possible that chaos theory will provide a viable mechanism for amplifying quantum effects into large-scale macrophysical results, and hence that chaos effects could be used by an omniscient being in combination with quantum indeterminacy to bring about its purposes. (The 'could' here should be emphasised: there is as of yet no established physical theory that brings together quantum mechanics and chaos theory in the fashion envisioned here.)

Finally, we found that a complete account of divine agency would have to include a 'top-down' component. I argued that Peacocke's conceptualisation of top-down agency in terms of emergent properties is preferable to Polkinghorne's somewhat too brief comments on top-down causality as information-based *rather than* energy-based. Still, both versions are correct in stressing that a theologically adequate account requires that we take account of the holistic properties of entire systems, which are not reducible to the (physical) components of these systems.

Perhaps most significant in our study of the topic was recognising a common argumentative move that cropped up in virtually all the recent articles on divine agency in light of contemporary science. Time and time again we found references in the texts to the analogy between, on the one hand, human consciousness (or mental properties) and the body and, on the other hand, the relationship of God to the world. It is time now to see whether we can reconstruct this analogy as a careful argument so as to obtain a systematic theology of divine agency.

IN DEFENCE OF EMERGENTIST SUPERVENIENCE

The thesis of this theory of divine agency can now be clearly stated: theology does not need to embrace *either* a radical dualism of mind/soul and body *or* the physicalism that is widespread among scientists and philosophers today. The theory of *emergent properties* forms an attractive *via media* between these two poles of the discussion. The attraction of the emergentist approach lies both in its adequacy as an overarching framework for relating various scientific disciplines and in its resources for specifying God's relation to the world. Rather than *reducing* 'life' or 'mind' to the lowest common denominator, this approach sees them as emerging out of the lower-order structures as a genuinely new type of phenomenon.

Emergence

Emergence has strong scientific credentials to back up its claim to be a conceptually coherent and empirically viable middle road between dualism and reductionism. The general principle of a scientific account of reality is, 'Take as real those postulated entities that give rise to the most powerful explanations of some set of data'.[20] From the atoms of Democritus to the quarks of recent research, the various entities postulated by scientific theories have depended on their explanatory successes for their primary justification. Note, however, that not all of these entities are fundamental particles of the lowest-level science, physics. The theory of emergence comes into its own because there are cases where explanatory success runs in the opposite direction – not down to the smallest parts but up to larger 'wholes' of which the parts are parts. C. D. Broad recognised a number of years ago that 'the characteristic behavior of a whole could not, even in theory, be deduced from the most complete knowledge of the behavior of its components, taken separately or in other combinations, and of their proportions and arrangements in this whole'.[21] As it turns out, better explanations are sometimes produced by regarding the whole as an existing entity in its own right. In such cases, we say that the higher-order entity – say, the cell or organism – has come into existence or 'emerged' from a certain composition of parts.

Although the era of logical positivism saw some resistance to the doctrine of emergent properties, there has been a revival of interest in emergentist hypotheses in recent years, especially in the field of the philosophy of mind.[22] In the article cited above, for example, O'Conner defines an emergent property as 'a (typically) simple, non-structural natural property that is exemplified by objects or systems that attain the appropriate level and kind of organizational complexity and that exerts a causal influence on the behavior of its possessor'.[23] Cells consist of complicated molecules, the properties of which are analysed by chemists; molecules are composed in turn of atoms, which fall under the domain of atomic physics. The language of biochemistry dominates research into cell behaviour. Yet the explanations of cell growth and reproduction are finally given not in chemical or physical terms, but in terms of the higher-order phenomenon itself, the cell. The cell is an emergent phenomenon, a distinct thing in the world with its own explanatory science – even while it is composed out of smaller units and its behaviour conforms to more foundational physical laws.

Beyond cells there are special disciplines devoted to the biology of individual organisms, which thus represent another level of emergence: plant physiology, zoology and the biological study of (say) higher primates. Each of these disciplines involves a unit of selection and thus requires an explanatory system in its own right. Especially in cases of the phylum Chordata (of which vertebrates are a subphylum), with its more highly developed nervous system, it is clear that the cellular level of analysis alone is not sufficient to express the causal interdependencies of the variety of behaviours of the organisms in question.[24] As Varela and his colleagues have written:

> There is no unified formal theory of emergent properties. It is clear, however, that emergent properties have been found across all domains – vortices and lasers, chemical oscillations, genetic networks, developmental patterns, population genetics, immune networks, ecology, and geophysics. What all these diverse phenomena have in common is that in each case a network gives rise to new properties, which researchers try to understand in all their generality. One of the most useful ways of capturing the emergent properties that these various systems have in common is through the notion of an 'attractor' in dynamical systems theory.[25]

What then of mental phenomena? In the case of complicated organisms like ourselves, we find it necessary to assert a type of predicate unlike any biological qualities previously encountered. The fact is that the neo-cortex, with its holistic states and high degree of interconnectedness, gives rise to phenomenal experiences that we call mental. Mental events are experienced as 'more than' anything that pertains to the organism or even to its brain; they appear to us as irreducible. How is this 'more than' to be thought? A major recent school of philosophy views these mental states as 'supervening' on physical states. As we will see in a moment, the supervenience concept is a more rigorous way of expressing what the doctrine of emergent properties expressed in a more rough and ready way: mental properties depend on their biological substratum (we do not ascribe mental predicates to corpses!), yet they are not reducible to that substratum.

So the theory of emergence is three things: a theory of explanatory adequacy; hence, a theory of causal activity; and finally – assuming that we can assume that scientific success tells us *something* about what exists – an ontology, a theory of what exists. The key observation here is that 'higher-order' entities – entities composed out of so-called foundational particles – sometimes also evidence higher-level forms of causality. As R. W. Sperry writes,

These micro-interactions and the interrelations of all the infra-structural components become embedded within, enveloped, and as a result are thereon moved and carried by the property dynamics of the larger overall system as a whole . . . that have their own irreducible higher-level forms of causal interaction.[26]

It is not that lower-level phenomena are simply placed in a broader context while otherwise remaining fully valid; rather, 'because an emergent determines (in large part) a relational complex that cannot be adequately described in terms of lower level components and their micro-relations, the micro-physical laws are inadequate for, cannot be applied to, such situations'.[27]

The Concept of Supervenience

As noted above, supervenience theory is a particularly intriguing development for theology.[28] Since it grants that a physical substratum, the brain, is a necessary condition for the supervening mental state, it does not stand opposed to the brain sciences and can draw on their positive conclusions (no 'soul of the gaps' strategy here); yet mental states are not reducible to physical processes. Like the theory of emergent properties, though in a more rigorous fashion, it holds the mental to be something different from the brain states upon which it rests.[29] It was the careful formulation of how consciousness could arise out of causes in the brain by thinkers like Sperry which led to supervenience theory:

Instead of following the usual approaches that tried to inject conscious effects into the already established chain of microcausation, the logical impasse was resolved by leaving the microcausation intact but embedding it within higher brain processes having subjective properties with their own higher-level type of causation, and by which the embedded micro-events are thereafter controlled . . .

[E]xcitation of a cortical cell is enjoined into the higher dynamics of passing patterns of cognitive activity. A train of thought with one mental thought evoking another depends throughout on its neurocellular physiology and biochemistry. Nevertheless, like molecules in passing waves in a liquid, the brain cell activity is subject to higher-level dynamics which determine the overall patterns of the neuronal firing, not relative to other events within this particular brain process, but relative to the rest of the organism and its surroundings.[30]

Supervenient properties, then, are properties that cannot be fully 'parsed' in the terms of one level but require the introduction of a second level, since they involve properties of a different kind or status. One way to think these supervenient properties is to use the language of *holism*; we can speak, for example, of the systemic properties of a particular system, properties that are not reducible to specific interactions, energy levels, and so on. Thus the boundary conditions of a physical system affect all the particles and energy distributions in that system; yet they cannot be reduced to such interactions alone. There is some analogy between such holistic or boundary conditions and the type of supervenience here envisaged.

In his influential paper, 'Supervenience and nomological incommen-surables',[31] Jaegwon Kim presented the more limited (or less metaphysical) version of supervenience. Very different sorts of properties can be related in lawlike ways, even when the *properties* are not reducible to one another. The properties of our (macrophysical) world, for example, certainly do not *seem* to be composed of fundamental particles or to obey the laws of quantum physics. Nonetheless, we hold that particle physics is the basic science and that processes at that level determine the behaviour of the large objects of our experience. Why should the same not be true of mental properties and the brain? The properties of mental experience are unlike brain states, and successful brain science need not reduce or eliminate them; yet still (Kim thought) they are the causal result of brain processes alone.

Kim's position has the merit of preserving the reality of mental states against reductive physicalists such as the Churchlands. The reductivist view suggests that the way in which we normally speak of the mental – thoughts, intentions, will, self-consciousness or what they call 'folk psychology' – will eventually be left behind, replaced by the terms of a (future) successful brain science. Supervenience at least avoids that fate. At the same time, in Kim's hands it does not really allow for the emergence of something ontologically new; the qualities of the mental life are preserved, but the level of the mental is not really ontologically distinct. Ultimately, it is causality at the physical level – the physical system as a whole, if you will – that fully accounts for the mental. Thus Kim would accept a test such as the one formulated by Eaton:

> TEST: If one can imagine two objects or events identical in all respects except that one has property P and the other does not have property P, then P is not supervenient; if one cannot imagine two objects or events identical in all respects except that one has property P and the other does not have property P, then P is supervenient.[32]

Emergent Supervenient Properties

Theologians – but not only they! – will recognise that this version of
supervenience is not strong enough. As it turns out, there are a variety
of types of emergent or 'holistic' properties. So, for instance, the boundary conditions in physical systems constrain, and in this sense influence,
physical states. Features of the system as a whole may include qualities
that do not pertain to the parts of the system taken on their own, and
they may even influence the behaviour of the parts of the system.
Temperature may be a supervenient quality that pertains to a very large
number of hydrogren and oxygen atoms but not to the atoms individually. The passing of a wave through water causes a movement of
molecules that must be explained by moving up and not down the
explanatory chain. The process of cell reproduction or the movements
made by an individual organism are likewise emergent properties; at the
same time, they are processes that influence the atoms and molecules
making up the cell or organism.

Not all emergent properties are equal. As Eaton points out, does not
a predicate such as 'is our leader' – which emerges only at the political
level but not at the individual level – supervene on a collection of
human beings in a *different* way than temperature? It does not seem
'connected to or determined by the microphysical level the way "is
boiling" or "is spherical" or even "wants to go to Nairobi" do'.[33] Aesthetic
and ethical or spiritual qualities would presumably diverge as well. It
therefore appears necessary to introduce a distinction between weak
supervenience (the view of Jaegwon Kim and his followers) and strong
supervenience. The only view that will be sufficient for theological
purposes is strong supervenience. Let us see what kind of a case we can
make for it.

The argument has three steps. The first is the premise that *mental
predicates supervene on physical states.* That is, mental properties arise out of
and are not reducible to physical properties; they cannot be fully
described in physical terms. At the same time, they seem to depend
upon their physical basis: when the brain stops functioning, mental states
do not continue (at least in any form *we* can observe!). Here Kim and
company are allies against those who defend the abolition of folk psychology, the eventual replacement of psychological language with talk of
physiological states. If even weak supervenience is correct, your mental
experiences can be taken really to exist; thus, for instance, your intro-
spective reports are not merely *mistaken* accounts of what are 'no more
than' neurological events.

So far, this is a minimal and relatively uncontroversial claim. It is to say only that in dealing with mental phenomena we witness the appearance of non-physical characteristics. It is not yet to say that anything other than physical causes are at work or that anything other than physical things exist. There may not be scientific reason to introduce a new kind of 'stuff' (e.g. mental substance or 'soul'), but there is sufficient reason, within the context of the natural world, to speak of a new *kind* of property. Thus we speak of these properties as emergent.

Second, *one supervening mental property must be said to cause another.* With this move the theologian breaks with the physicalist position of someone like Jaegwon Kim. As a physicalist, Kim must insist that all causation is physical causation. In the process of criticising mental causation, he offers a concise and clear summary of the emergent supervenient hypothesis – the hypothesis that, I will argue, offers the strongest overall position on body and mind. This view grants that mental phenomena depend on their physical substratum, while insisting that new types of causal activity emerge. As Kim describes the position:

> The thesis of Property Emergence is the emergentist expression of this idea of psychophysical dependence. It says that mental phenomena emerge when, and only when, certain appropriate conditions are satisfied by the biology of the organism; moreover, mental phenomena must, as a matter of law, emerge when the right conditions are present. This is very much the picture we get from the talk of the 'physical realization' of mental states . . . Thus, corresponding to the thesis of property emergence is the nonreductive physicalists's claim that mental properties are instantiated only by being realized by physical properties in physical systems.[34]

Specifically, Kim realises that any philosophy of emergence 'stands or falls' with the notion of downward causation. This is correct: the emergence view argues for the real existence of the resultant properties. Now 'to be real is to have causal powers'.[35] For example, we speak of the causal powers of cells, organs and brains because doing so provides the best explanations of the biological data, and *because* these theoretical terms have explanatory power we consider them to exist. The key question then becomes: do brains have holistic or emergent properties that are unique – properties that we must say are essentially mental rather than physical? I argue the affirmative.

Note that to argue in this fashion is not the same as to posit a mental 'thing' or 'container' in which these mental properties reside. One reason for caution here is that the causal relations between this mental 'thing' and the brain with which it is connected would become difficult, if not impossible, to conceive. The unsolved problems of mind/body interaction that were raised by Descartes' dualism continue to militate against theologies that work with a 'soul-stuff'.[36] Emergent supervenience allows for new qualities to arise out of an existing object like a brain, but it does not defend the introduction of a kind of substance in which these thoughts reside or 'inhere' (call it the mental milk bucket view). We need not divide humans into two fundamentally separate kinds of 'things'. Instead, thoughts can be taken as new properties that supervene on a physical system that we already know and can study.

Take *weak supervenience* to be the view that it is sometimes justified to postulate mental properties without having to reduce them to their physical (subvenient) basis (e.g. Kim's position).[37] *Strong supervenience* would then be the view that these mental predicates can also have causal effects, effects not reducible to the physical events occurring in the brain. Using the vocabulary of emergent 'levels' (say, L_1, L_2, L_3, etc.), one can say that biological explanations represent L_2 explanations which are not reducible to explanations at the level of the physical systems (L_1) out of which they arise. *Strong supervenience* holds that mental (L_3) properties, states and phenomena supervene on L_2 if and only if (1) L_3 emerges out of L_2 and remains dependent on it for its existence, yet (2) L_3 evidences causal interactions that are *sui generis* and not reducible to causal interactions at the level L_2. Kim summarises the 'heart' of this doctrine of mental causation: 'Mentality must contribute genuinely new causal powers to the world – that is, it must have causal powers not had by any physical-biological properties, not even by those from which it has emerged.'[38]

To common sense or 'folk psychology', the causation among ideas seems obvious. You have the idea of '23 + 47' and that idea gives rise to the idea of '70'; the causation in question goes from the one idea to the other. It is not explanatorily equivalent to describe this causal sequence in neurophysiological terms alone: you are in brain-state C451 and that, together with a variety of other neurological conditions, gives rise to brain-state D773, which happens to have the supervening property of being the idea of 70. It is unnecessary to make the long detour of forming a one-to-one correspondence between the first idea and a particular physical state which then gives rise to a second physical state which in turn is perceived by the individual person as the idea '70'. It is more natural to begin with the sequence of causal connections at the intellectual level – a sequence that we speak of as 'one idea giving rise to another'. We do not yet know whether, in describing the neurophysio-logical underpinnings of this process, it will turn out that there is indeed a one-to-one correspondence between ideas and particular neuronal fir-ings, or whether the process of thought works in a much more holistic fashion.

Against the opponents of mental causation, I argue that *as long as* the explanatory power of idea–idea causation continues to be much greater than the neurophysiological account, we should straightforwardly assert its superiority and indispensability. Indeed, we have good reason to think that even an advanced neurological science, successful beyond the wildest imaginings of the person on the street, would not provide the simpler causal account – any more than a more advanced physics would lead us to tell the story of cell reproduction in terms of physics alone. In both cases the mathematics would be incredibly complex and the predictive value close to nil.[39]

We might put the point schematically (borrowing from yet modify-ing Kim's more critical treatment). Let us begin with a certain physical property or state P, say a holistic brain state. Some mental property M supervenes on P, say the mental property *thinking 23 + 47*. The claim is that M can cause another idea or mental state, $M^\star$ – in this case, say, the *idea of 70*. According to supervenience theorists, $M^\star$ can obtain only because some physical property or state (call it $P^\star$) obtains. The question is, should the causal line be drawn as:

1.

$$M \qquad M^\star$$
$$\uparrow \qquad \uparrow$$
$$P \longrightarrow P^\star$$

or should it be drawn as:

2.

Actually, there is another possibility, which constitutes the final step in the present three-step argument; it is the claim that *mental properties are not epiphenomenal but can in turn cause physical events*. Again, for common sense this is not a problem at all. Presumably every reader believes that he or she can form the (mental) intention to cease reading this book (and perhaps most are by now wishing that they could!). That (mental) idea could in turn cause your eyes to cease moving across the page – a physical result. True, many philosophers of mind resist this conclusion; yet

I suggest that the explanatory power of mental ⟶ physical causation is greater than those views that deny it. The result might look like this:

3.

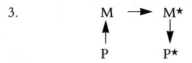

Here a brain state, $P^\star$, is caused by a mental predicate. This would be a genuine case of 'downward causation'. Ideas can cause other ideas, and ideas can cause changes in the physical world, for instance my lifting my hand, my travelling to China, my writing a book. Even Kim admits that the enduring appeal of the emergentist view of mental causation stems from its combination of two notions:

> The idea that mentality emerges out of, and in that sense depends on, the physical, and the idea that, in spite of this ontological dependence, *it begins to lead a causal life of its own*, with a capacity to influence that which sustains its very existence – that is, the combination of 'upward determination' and 'downward causation'.[40]

Downward causation marks a major parting of the ways, however. Even those physicalists who grant (weak) supervenience must find downward causation a problematic belief, since their account looks more like picture (1) above: 'The only way to cause an emergent property to be instantiated is by causing its emergence base property to be instantiated.'[41] I have maintained, by contrast, that mental causation makes the best sense of the phenomena of human experience. That it is the only position that would allow for the action of God upon human minds is certainly of great significance for theologians. Yet I have made the case not from theological grounds alone but rather from more general considerations about the best explanation of human thought and behaviour. That is, real mental–mental causation is a position that observes the canons of scientific method and heeds developments in neurophysiology. The challenge to reductive materialism, as well as to a physicalism that makes mental properties epiphenomenal rather than causally active, comes from within the disciplines in question and not from the outside.

To some theologians it may seem like too much to relinquish the traditional beliefs: that there are mental *substances*, that mental properties inhere in mental substances, that humans are fundamentally divided into separate kinds of 'things'. But the explanatory power of classical dualism is not great enough to justify its costs, among which is the still unsolved problem of soul–body interaction (remember Descartes' futile attempt to locate the point of intersection between the human soul and the human body in the pineal gland, whose function otherwise eluded him). Against many theorists today, it is still possible to move from neurophysiology to talk of consciousness.[42] Yet consciousness cannot be a metaphysical surd; it must be in some way analogous to structures encountered elsewhere in the study of the world. The framework of emergence is sufficient to establish this analogy.

Emergent supervenience may seem overambitious to physicalists, overcautious to theologians. The caution is fully justified, however. These are about the strongest claims that can be made *and taken seriously* within the contemporary science-oriented discussion. Given the evidence, one *can* say that a 'strange' type of property supervenes on physical systems and that these mental properties constrain (and in this sense act causally upon) the system in question. Anything beyond this point is, to our scientific discussion partners, metaphysics in the bad sense. For example, although the person on the street might be tempted to speak of mental substances – the 'mind', 'inner self' or 'soul' – that do the thinking and thereby manifest themselves in the world, such extrapolations from the data have no place within science. Even the contention that supervenient properties are 'emergent', that they represent an order of reality different from that out of which they arise, pushes at the limits of what one can justifiably say. Unlike dualism, however, emergentism can be defended by its explanatory value.

THE DIVINE SUBJECT: THEOLOGY'S UNIQUE CONTRIBUTION

Let us summarise the argument to this point. I first defended the theory of emergence. Physical systems of a sufficient degree of complexity evidence 'holistic' properties not reducible to the direct causal influence of any of their constituent parts. Using the vocabulary of emergent 'levels' (say, L_1, L_2, L_3), I argued that biological explanations are L_2 explanations that are not reducible to the physical (L_1) systems out of which they arise. Using arguments drawn from the biology/physics relation, as well as from folk psychology, the experience of human agency and the recent philosophical discussion, I argued that emergent properties *supervene* on the properties and laws of the constituent parts. I then made the case for *strong supervenience*: mental properties evidence causal interactions on a third level (L_3); they are not reducible to causal interactions at the biological (L_2) or physical (L_1) level. That is, mental properties can also be the direct cause of other mental properties as well as the cause of changes in the physical world (e.g. in one's own brain).

At first blush, this entire discussion would seem to be bad news for theology. If mental phenomena supervene on physical systems such as the brain, would this not make the existence of the mental realm intrinsically dependent on the physical? Would that not in turn make it impossible for there to be disembodied existence – or existence in some non-physical state after death in which the human person continues without a body? If the human brain is the necessary basis for any mental functioning, it would seem arbitrary to posit any mental functioning after the death of the body, whether through postulation of a 'new heavenly body' or not. If that be the conclusion, gone is the hope of eternal life – and gone too is the credibility of belief in God as a being who also transcends the universe and could continue to exist without it. Supervenience may one day win as the best position in the philosophy of mind, but it might seem that, if it does, its victory will be of no more help to the Christian theologian than reductionism itself.

Note that two possible ways out of the dilemma are closed to us. One cannot return to a Cartesian dualism if it makes mind/body interaction more of a mystery than it was before. Nor can one break all links between human and divine agency, conceding the field to the philosophers of mind and simply insisting that God is an agent in a way that has no connection whatsoever with human agency. For, as we discovered, if we do *not* conceive of divine action on analogy with human agency, we will have no empirical control, even of an indirect sort, on the theology of divine agency. Divine agency will either become a pure mystery (fideism) or an arbitrary matter to be 'metaphysicised' as each particular author or theological tradition pleases.

Theology and the Doctrine of God

In order to see the path to a solution, recall the theological argument to this point. We found strong theological reasons to maintain an analogy between the human and the divine agent, and hence between human and divine agency. In previous chapters we developed the analogy in some detail: just as human consciousness (mental properties and their causal effects) can lead to changes in the physical world, so also a divine agent could bring about changes in the physical world – if this agent were related to the world in a way analogous to the relationship of our 'minds' to our bodies. With this analogy we found ourselves led again to the theological position developed in earlier chapters of this study: panentheism. A panentheistic account of the God/world relation provides, I argued, the best framework for a theory of divine agency.

What does one do when the scientific results seem to point in a direction that is at odds with fundamental theological tenets? If one is involved in natural theology in the traditional sense – moving inductively from scientific results to construct one's theology – then one must draw the consequences and live with them. But this is not the model that we found ourselves adopting in the discussions in Chapters 5 and 7. There we found the scientific results to *underdetermine* one's choice among metaphysical interpretations. Authors involved in meta-scientific debates inevitably supplement the science with a metaphysical perspective of their own – even when they label their arguments 'purely inductive'. Of course, this fact is no justification for ignoring the scientific results. But it does show that the science/metaphysics relation is simply not so 'tight' that all appeal to metaphysical considerations is eliminated or subordinated.

The role of metaphysical arguments is increased as one moves further away from empirical questions. On questions within 'metaphysics proper', considerations of consistency, systematicity, coherence and comprehensiveness play the guiding role; empirical data become illustrative rather than controlling. So, for example, when John Leslie considers multiple cosmologies in *Universes*, his argument is much more constrained by theoretical physics than when he argues for an innate value in the universe in *Value and Existence*.[43] The latter position, however strongly it might be suggested by universal evolution, *must* depend on some fundamental intuitions about the value-ladenness of the universe.

What pertains to metaphysics pertains to systematic theology as well. Beliefs about the nature of God prior to and apart from the universe can never be dictated by anything *within* the universe, although what we know of the world will of course influence what we believe about God. If a theological belief involves claims that never could be empirically tested, then we cannot fault the belief for not providing empirical warrant.[44] This fact is of enormous significance, for it allows theology to enter into the debate with the sciences as an equal partner *at those points where the discussion concerns matters that are not (and could not be) empirically resolvable*. Theologians must be careful about claiming 'metaphysical immunity', since the realm of empirical control has proven to be much broader than was once thought; it is all too easy to fall into a 'God of the gaps' approach to theology. It thus behoves us to pay careful attention to the point at which we cross over the line into empirical debates – which is not to say that we ought never to cross it.

The issue at hand – does God essentially depend on the universe, or is he in some sense independent of it? – is a paradigmatic case of a trans-empirical question. Nothing within the world could dictate the answer to this question, since nothing within the world could determine whether its source is essentially independent of it. This means that we now have a theological (or metaphysical) reason to insist on a major *dis-analogy* between God and agents within the world (at least in so far as we know them through empirical study). God's being, and any qualities that accompany it as part of God's eternal nature, precede the set of experiences that God might have within the world. This is a metaphysical move; theologians postulate God as a being of its own and not merely as an emergent set of divine properties. Moreover, we postulate that God is not dependent on the world for existence but preceded the world and created it. Finally, we postulate theological properties (eternality, omnipotence, moral perfection) that are not and could not be derived from the analogy with human agents. These (metaphysical or theological) beliefs are not and could not be inferences from the physical world. Yet they are, I have argued, fully allowable in terms of what we know about the physical world, and they might even be said to be *suggested*, at least in part, by what we know of that world.

The *theological supplementation of science* has important implications for our formulation of panentheism as well. It does not eliminate the important biblical and philosophical arguments canvassed in previous chapters on behalf of the intimate link between God and world. But it does bring panentheism more into the spirit of biblical theism, in which God's transcendence of the world is never cast into question by the recognition of the intimate dependence of the world on God. Moltmann's panentheism, for example, is characterised by an embracing of process and immanence language about God *at the same time* that God remains a centre of agency apart from the world. There are a variety of ways to give metaphysical expression to this moment of difference: the immanent and economic trinity, the antecedent and consequent natures of God (process theology), potency and actuality.[45] The important thing is that biblical texts and theological reflection may be used to supply further content about the nature and intentions of God. The panentheistic analogy provides a guiding framework, but this does not mean that it is capable of serving as the sole source for one's doctrine of God.

Following the Panentheistic Analogy in the Other Direction

If theological considerations can supplement what we know of God's nature from the world, can they not supplement what we know about human nature as well? We found that a full account of God involved talk of God's being apart from the world's existence. Even if all of God's *experiences* were derived from the world – a conclusion vehemently denied by any trinitarian doctrine of God – there could still be grounds for speaking of God's self as more than the sum of these experiences. Could the same be true of humans? That is, could *what it is to be a human agent* involve more than the set of mental properties – thoughts, wishes and intentions – to which we have empirical access?

With this possibility we come to the third and final movement in this book's argument. First, in an upward movement of reflection we used the panentheistic analogy to supply content and guidelines for our understanding of what it means to speak of God as an agent at all. Then, in the previous section (as in Chapters 5 and 7), we supplemented what could be learned from the analogy with additional information from metaphysical and theological reflection. Now I am suggesting that we can follow the panentheistic analogy back down, moving in the other direction given what we have learned through the first two types of reflection. (Of course, in another sense the downward movement is not new at all, since the book opened with an extended discussion of the God/world relation in the Hebrew Bible and New Testament.) In other words, once one has reflected theologically on the nature of God, one is justified in turning back to science with these results in hand. It is now possible to provide a theological interpretation – and even correction – of the results derived from science alone.

For example, science could never justify speaking of a mental entity, 'the subject', which is the possessor of mental properties. But the theological account of divine agency reveals a divine subject which is both in the world (the world is analogous to God's body) and transcends it. Likewise, what we know theologically to be the nature of the divine subject we may postulate (through a theological inference, and *mutatis mutandis*) to pertain to the human subject as well. Here we have the classical theological move: the use of the *imago dei* as a source for the interpretation of human nature. Both are characterised, for example, by rationality, morality and eternality. Also, as we learn from the doctrine of the Trinity, both are essentially social or relational in nature.[46]

The downward panentheistic analogy also allows us to speak of the *agent* who possesses the mental properties and who exercises the mental causation that I defended above. The use of a metaphysics of the subject is of course not new or unique to theology. Plato, Aristotle, Descartes, Leibniz, Kant – all of these thinkers supplemented the direct data of human experience with their own notions of the metaphysical or transcendental subject that underlay these experiences. It seems clear that *some* such account of the thinker is necessary even in light of the best work in neurophysiology and the philosophy of mind today.[47] It has been the particular merit of the work of Wolfhart Pannenberg to demonstrate how the reflection on the nature of the human subject in German Idealism provides perhaps the most sophisticated acount of the conscious subject.[48] A theory of thoughts (mental predicates) that does not include reference to the active, conscious subject whose thoughts they are is inherently incomplete.[49] Here an especially strong case can be made that the theological account of the human agent provides a needed supplement to recent work in the brain sciences and the philosophy of mind. As a result, we have theological reason to interpret talk of mental causation (mental to mental and mental to physical) in terms of the actions of individual human subjects.

The Break with Materialism

What about the soul, then? Does it fall within the category of allowable inferences from the divine nature? On the one hand, one does not sense a large jump from the theory of human agents just described to the classical doctrine of the soul. If the move is made, one can speak (theologically) of eternal but finite subjects, created by God as finite versions of divine personal being and guaranteed an eternal existence by their very nature. On the other hand, whereas the divine (threefold) subjectivity is eternal and dependent on nothing outside itself, human subjects are essentially contingent and dependent upon God. So there is also theological reason to resist postulating an essentially eternal soul as underlying human experience. Even without a substantial soul, however, one can assert that God by his grace preserves the existence of the subject after the death of her body for eternal fellowship with God. Thus in either case the biblical hope of a future resurrection and life in the presence of God is preserved.

Once we are allowed to follow the panentheistic analogy in both directions, a rich interchange of scientific and theological intuitions begins to emerge. As Jantzen has argued, the assertion of panentheism is not that God has *a* body like us, but that he stands in a relation to *the universe as a whole* that is analogous to your relationship to your body.[50]

You are 'actualised' in your body; it contributes incredibly to your sense of what the world is, whether it is good or not, what you know about the world, and how you communicate your most unique ideas to others. I cannot know you *apart from* your body; I infer that you are happy when you smile warmly and that you are sad when you weep; you please (or anger) others by touching their bodies, and they affect you by words and expressions that are mediated through their bodies. It is not correct to reduce you to your body, to say that you are nothing more than it. But it would also not be correct to separate you too greatly from your body, as if 'you' were primarily and in essence a purely spiritual 'thing' which just happened to be inhabiting that particular human body, say in the way that a late-night driver checks into a motel for a few hours of sleep before continuing on her journey. Such dualistic ways of thinking of the human person are no longer credible; 'Indeed,' writes Jantzen, 'it sounds quaint to our post-Darwinian ears to hear talk of souls using our bodies to get from place to place or to accomplish their activities, as though bodies were a kind of lorry for the transportation of souls.'[51]

As Jantzen notes, the necessity of distinguishing God from the world was based in no small part on the conclusion of Plato and Aristotle that matter was inferior to form – less real, less perfect. A perfect being (*ens perfectissimum*) would therefore have to be pure form, not sullied by matter in any way. Now one might speculate about the philosophical adequacy of this view. But one development in science certainly has something to say about it: the convertibility of matter and energy postulated in Einstein's famous equation (the one equation that even those who have no interest in mathematical physics seem to know), $E = mc^2$. If there is *any* scientific falsification of metaphysical statements, this is certainly a good example: matter and form or energy cannot be distinct metaphysical principles or building blocks of the world if they are inter-convertible and in fact a continual flux between the two takes place (under certain conditions).

This insufficiency is even clearer if theoretical work on forcefields, and particularly work toward a unified field theory, is correct, since in this case physical things will not be defined primarily in terms of matter at all. Classical thinkers have been happy to connect God with units of energy within the world, either as the ultimate source of all motion (Aristotle, St Thomas) or as the highest of all mental entities and the source of them all (Plato, Descartes, Leibniz). Direct lines of implication can be drawn from these views to (*inter alia*) Augustine's theory of the work of God's Spirit in the world, the divine light of Calvin, and

Luther's distinction between nature and grace. But Augustine, Luther and Calvin did not move on to draw similar connections between God and matter in the world. What happens if the spirit/matter distinction is rethought, as appears to be necessary today? I have argued, following Grace Jantzen and others, that redrawing the lines will force us to think much more carefully about the metaphor of the world as God's body than the tradition has done heretofore. As she writes:

> The idea of matter as utterly alien to us, incapable of thought or life, and hence unworthy to be predicated of God, is a view which theology must follow science in rejecting. The holism and integration of human persons which religion strives for, integration with the world, with other men and women, and with our own bodies, is incompatible with a derogatory evaluation of matter. But this means that, with a more positive view of matter, the way is open to consider its relationship to God, on the model of divine embodiment in the world.[52]

Conclusion

With the addition of the second 'direction' in the panentheistic analogy we have completed the argument of this book. If the argument is correct, we have achieved what we sought for: to formulate a theory of God as divine agent which is both a product of theological reflection *and* consistent with (and perhaps even suggested by) what science has come to know about the natural world and the place of human agents within it. According to the panentheism I have defended, God can act on any part of the world in a way similar to our action on our bodies. At the same time, God also transcends the world and will exist long after the physical universe has ceased (or has died the death of entropy).

This theological conclusion grounds our hope that God will break the limits of the mind's dependence on the body – a dependence that is ubiquitous in our present experience of the world. The Christian hope is that, by his grace, God will enable the continued existence of the self in conjunction with a non-physical body after the death of our physical bodies. It also leads us to hope that, although the universe shall surely be uninhabitable for human subjects at some point in the finite future, God will create 'a new heaven and a new earth' in which human subjects can dwell eternally in the divine presence.

Our examination of the theology of divine agency and God's relation to the world in these pages has been characterised by careful attention to the methods and results of science. The panentheistic analogy, on which important parts of the argument have relied, represents the attempt to move from our knowledge of human agency toward an understanding of what the theological assertion means: that the God of the Hebrew and Greek scriptures is an agent in some (limited) ways like human agents. Let there be no doubt about it, however: in these final sections we have moved beyond the realm of empirical knowledge and control. Christian beliefs about the transcendence of God and God's ultimate plans for this universe (and beyond) are not 'checkable' by any scientific means – though they will be subject to an ultimate or 'eschatological' verification or falsification at the end of history. I have sought for a faith that is as informed and as critical as it is possible to be. Still, the argument has made clear that theological assertions never fully divest themselves of a component of faith and trust. In retrospect, was this fact not anticipated already in the opening chapters, where the biblical materials set the context for the philosophical and scientific reflection that followed? The transcendence of God can no more be 'read off' the world than the soul can be proven from the existence of mental phenomena or the eschaton demonstrated on the basis of our knowledge of physical cosmology. The combination of science and theology, however, does provide a rich source in our quest to know 'in whom we have our being'.

NOTES

1. Colin McGinn, 'Can we solve the mind-body problem?' in Richard Warner and Tadeusz Szubka (eds), *The Mind-Body Problem: A Guide to the Current Debate* (Cambridge: Blackwell, 1994), pp. 113f.
2. Schubert Ogden, 'What sense does it make to say, "God acts in history"?' in Owen C. Thomas (ed.), *God's Activity in the World: The Contemporary Problem* (Chico, CA: Scholars Press, 1983), p. 89.
3. Charles Taliaferro, *Consciousness and the Mind of God* (Cambridge: Cambridge University Press, 1994), p. 248.
4. Thomas Aquinas, *Summa Theologiae*, I/I, q. 93, art. 3.
5. See, for example, Grace Jantzen, *God's World, God's Body* (Philadelphia: Westminster Press, 1984); Elaine Pagels, *The Gnostic Gospels* (New York: Vintage Books, 1989).
6. Robert H. Gundry, *Soma in Biblical Theology: With Emphasis on Pauline Anthropology* (Cambridge: Cambridge University Press, 1976).
7. Jantzen, *God's World, God's Body*, p. 4.
8. See, *inter alia*, Paul Churchland, *The Engine of Reason, the Seat of the Soul*

(Cambridge, MA: MIT Press, 1995); Daniel Dennett, *Consciousness Explained* (Boston: Little, Brown, 1991); Antonio R. Damasio, *Descartes' Error* (New York: G. P. Putnam's Sons, 1994). It may well be that many of these works make materialist assumptions that set them at odds with theological reflection. Still, they bring extensive arguments for a more physicalist theory of mind that cannot be ignored and that demand our closest attention and our best responses.

9. This was the position taken by Edward Harrison in Chapter 5 above.

10. See John Leslie, *Value and Existence* (Oxford: Blackwell, 1979); and Leslie, 'Modern cosmology and the creation of life', in Ernan McMullin (ed.), *Evolution and Creation* (Notre Dame, IN: University of Notre Dame Press, 1985).

11. Jantzen, *God's World, God's Body*, p. 103.

12. Colin Gunton, *Becoming and Being: The Doctrine of God in Charles Harts-horne and Karl Barth* (Oxford: Oxford University Press, 1978), p. 2.

13. See the two important works by Joseph A. Bracken: *The Divine Matrix: Creativity as Link Between East and West* (Maryknoll, NY: Orbis Books, 1995), and *Society and Spirit: A Trinitarian Cosmology* (Selinsgrove: Susque-hanna University Press, 1991).

14. The partners Churchland remain influential spokespersons for this view. See, for example, Paul Churchland, *The Engine of Reason, the Seat of the Soul*, cited above, and Patricia Churchland, *Neurophilosophy: Toward a Unified Science of the Mind-Brain* (Cambridge, MA: MIT Press, 1986), as well as Patricia Churchland and Terrance J. Sejnowski, *The Computational Brain* (Cambridge, MA: MIT Press, 1992).

15. See Thomas Nagel, 'What it's like to be a bat', in Ned Block (ed.), *Readings in the Philosophy of Psychology*, Vol. 1 (Cambridge, MA: Harvard University Press, 1980), pp. 159–71.

16. See especially Paul Churchland, *A Neurocomputational Perspective: The Nature of Mind and the Structure of Science* (Cambridge, MA: MIT Press, 1989); and Paul Churchland, *Matter and Consciousness: A Contemporary Introduction to the Philosophy of Mind*, revised edn (Cambridge, MA: MIT Press, 1988).

17. See Thomas Nagel, *The Structure of Science: Problems in the Logic of Scientific Explanation* (London: Routledge & Kegan Paul, 1961), and Nagel, *The View from Nowhere* (New York: Oxford University Press, 1986). See also Frank Jackson, 'Epiphenomenal qualia', *Philosophical Quarterly* 32 (1982): 127–36, and a number of the articles contained in Steven J. Wagner and Richard Warner (eds), *Naturalism: A Critical Approach* (Notre Dame, IN: University of Notre Dame Press, 1993).

18. See Ned Block, 'What is functionalism?', in Block (ed.), *Readings*, Vol. 1, cited above.

19. Farrer's work on this subject is often reduced to his advocacy of double agency, a view we discussed in the previous chapter. But in fact underly-ing his concrete theory of divine causality is a richer account of God's relation to the world. When Farrer asks in *Faith and Speculation* (London:

Adam & Charles Black, 1967) about 'how to conceive the joint (as it were) between the action of the universal Will, and the actions of finite agents whether voluntary or merely natural,' he responds:

> We could scarcely do otherwise than follow the model our own being provides – a relation between a rational organic agency and the cellular agencies it organizes. [He later adds:] God is the Mind *of the World*, that is, he is not tied to any base of operation that is exclusively his; he enters fully into the subjectivity of all the world's constituents.

Now Farrer is no panentheist; his First Cause is transcendent of the world though immanent to its creatures, and Farrer explicitly rejects the metaphor of 'embodiment' (p. 158).

20. Cf. Bas van Fraassen, *The Scientific Image* (Oxford: Clarendon Press, 1980). Note that 'take as real' does not mean that one claims they are part of ultimate reality or some such. It means merely that these are the types of entities that, from a scientific perspective, we are most justified in postulating.

21. C. D. Broad, *The Mind and its Place in Nature* (London: Routledge & Kegan Paul, 1925), cited in Timothy O'Conner, 'Emergent properties', *American Philosophical Quarterly* 31 (1994): 91–105.

22. See O'Conner, 'Emergent properties', p. 91.

23. O'Conner, 'Emergent properties', p. 95.

24. Note that talk of emergent properties does not automatically sanction just *any* extrapolation; one is justified in introducing a science of organisms, or species, or ecosystems, or even the earth's biosphere as a whole, as in Lovelock's 'Gaia hypothesis', *only if* the new level of analysis provides better explanations and predictions than those given in terms of its constituent parts.

25. Francisco J. Varela, Evan Thompson and Eleanor Rosch, *The Embodied Mind: Cognitive Science and Human Experience* (Cambridge, MA: MIT Press, 1991), p. 88.

26. Roger W. Sperry, *Science and Moral Priority: Merging Mind, Brain, and Human Values* (Oxford: Blackwell, 1983).

27. See O'Connor, 'Emergent properties', p. 102.

28. In the science/theology dialogue, Nancey Murphy has made important use of the supervenience concept. See her *Beyond Liberalism and Fundamentalism: How Modern and Postmodern Philosophy Set the Theological Agenda* (Valley Forge, PA: Trinity Press International, 1996); and *Anglo-American Postmodernity: Philosophical Perspectives on Science, Religion, and Ethics* (Boulder, CO: Westview Press, 1997). The differences in our approaches will be clear to those familiar with these two books.

29. Thus John Bigelow writes that 'truth is supervenient on being' in *The Reality of Numbers* (Oxford: Oxford University Press, 1988), pp. 132–3 and 158–9. David Lewis predicates his own influential work upon Bigelow's guiding principle; see, for example, Lewis, 'Symposium: Chance and credence – Humean supervenience debugged', *Mind* 103 (October 1994): 473–89.

30. See Roger Sperry, *Science and Moral Priority*, pp. 231–2, quoted in O'Connor, 'Emergent properties', p. 102.
31. Jaegwon Kim, 'Supervenience and nomological incommensurables', *American Philosophical Quarterly* 15 (1978): 149–51.
32. Marcia Muelder Eaton, 'The intrinsic, non-supervenient nature of aesthetic properties', *The Journal of Aesthetics and Art Criticism* 52 (1994):385.
33. The example is taken from Eaton, 'The intrinsic non-supervenient nature of aesthetic properties', p. 385.
34. See Jaegwon Kim, '"Downward causation" in emergentism and non-reductive physicalism', in Ansgar Beckermann, Hans Flohr and Jaegwon Kim (eds), *Emergence or Reduction? Essays on the Prospects of Nonreductive Physicalism* (Berlin: W. de Gruyter, 1992), pp. 131f.
35. See Kim, 'Downward causation', p.135.
36. See the above cited work by Damasio, *Descartes' Error.*
37. Those familiar with Kim's position should note that I use 'strong' and 'weak' in the opposite sense from Kim. Kim speaks of 'strong supervenience' as complete determination of mental events by their physical substratum. He admits that 'supervenience of this strength entails the possibility of reducing the supervenient to the subvenient' (Jaegwon Kim, 'The myth of non-reductive materialism', *Proceedings and Addresses of the American Philosophical Association* 63 (1989): 46) – thus reducing the mental to the physical. As Marras has noted, a position this 'strong' (in Kim's sense) threatens to reduce mental phenomena to mere epiphenomena; see Ausonio Marras, 'Nonreductive materialism and mental causation', *Canadian Journal of Philosophy* 24 (1994).
38. Kim, 'Downward causation', p. 135.
39. See D. M. Armstrong, 'The nature of mind', in Block (ed.), *Readings in Philosophy of Psychology.*
40. Kim, 'Downward causation', p. 137.
41. Kim, 'Downward causation', p. 136.
42. In a similar vein, Hans Flohr has argued that the brain's subjective states can be causally potent:

> If subjectivity emerges from a system's representational and metarepresentational activity, it will be possible to describe these activities – the formation of representations and their interactions – in a physical framework. It will be possible to translate events in alien representational systems into activities of our own representational system. We would learn to manipulate tokens that represent representations in other representational systems. In doing so, we would not exactly feel like bats (or another alien system) ourselves, but we would be able to have a theory of what it is like to be a bat, i.e., we would know it . . . It also follows from the present hypothesis that the intentional contents of mental states likewise do not represent an epiphenomenon, but are real, causally effective states . . .'

(Hans Flohr, 'Qualia and brain processes', in Beckermann, Flohr and Kim (eds), *Emergence or Reduction?*, pp. 235, 234).

43. See John Leslie, *Universes* (New York: Routledge, 1989), and the references to his *Value and Existence* and other works cited in note 10 above.

44. It might be faulted for being cognitively meaningless, since it cannot be controlled by any possible empirical result. This was the contention of Anthony Flew and others in the famous 'University debate' (contained, e.g., in Baruch Brody (ed.), *Readings in the Philosophy of Religion: An Analytic Approach*, 1st edn. (Englewood Cliffs, NJ: Prentice-Hall, 1974), pp. 308ff.). I assume that Flew's objection has been adequately dealt with. Also note that it is not a protest against theology *per se* but actually against *all* forms of trans-empirical metaphysics.

45. In *Das Gottesproblem: Gott und Unendlichkeit in der neuzeitlichen Philosophie*, Vol. 1 (Paderborn: Ferdinand Schöningh Verlag, 1996), I described and (in part) defended this latter notion as developed in Schelling's 'metaphysics of freedom' (see especially Chapter 9).

46. See the detailed argument in Joseph Bracken, *The Divine Matrix*, cited above.

47. I have just seen David J. Chalmers' excellent new work, *The Conscious Mind: In Search of a Fundamental Theory* (New York: Oxford University Press, 1996), which shows the necessity of using supervenience theories to raise more fundamental questions about the nature of conscious experience. Thomas Nagel's publications, cited above, provided a further example of the need for this type of reflection.

48. See Wolfhart Pannenberg, *Anthropology in Theological Perspective*, trans. Matthew J. O'Connell (Philadelphia: Westminster, 1985), and the essays in *Metaphysics and the Idea of God*, trans. Philip Clayton (Grand Rapids, MI: Eerdmans, 1990).

49. See Charles Taliaferro, *Consciousness and the Mind of God*.

50. See Jantzen, *God's World, God's Body*, pp. 123f.

51. Jantzen, *God's World, God's Body*, p. 119.

52. Jantzen, *God's World, God's Body*, p. 122.

INDEX